LEGAL AND ETHICAL CONCEPTS IN ENGINEERING

KEITH W. BLINN

School of Engineering
University of California (Irvine)

Director
Energy Studies Program
Law Center
University of Houston

PRENTICE HALL, Englewood Cliffs, New Jersey 07632

Library of Congress Cataloging-in-Publication Data

Blinn, Keith W.
    Legal and ethical concepts in engineering / Keith W. Blinn,
        p. cm.
    Includes index.
    ISBN 0-13-528217-9
    1. Engineering law—United States. 2. Engineering ethics.
I. Title
KF2928.B57   1989
343.73'07862—dc19
[347.3037862]   88-39413
                        CIP

*To Deborah, Camille,
Erika, and Hadley*

Editorial/production supervision and
    interior design: Kathleen Schiaparelli
Cover design: Lundgren Graphics, Ltd.
Manufacturing buyer: Mary Noonan

© 1989 by Prentice-Hall, Inc.
A Division of Simon & Schuster
Englewood Cliffs, New Jersey 07632

Printed in the United States of America

10 9 8 7 6 5 4 3 2 1

ISBN 0-13-528217-9

Prentice-Hall International (UK) Limited, *London*
Prentice-Hall of Australia Pty. Limited, *Sydney*
Prentice-Hall Canada Inc., *Toronto*
Prentice-Hall Hispanoamericana, S.A., *Mexico*
Prentice-Hall of India Private Limited, *New Delhi*
Prentice-Hall of Japan, Inc., *Tokyo*
Simon & Schuster Asia Pte. Ltd., *Singapore*
Editora Prentice-Hall do Brasil, Ltda, *Rio de Janeiro*

CONTENTS

PREFACE

The primary goal of this book is to provide engineering students with some knowledge of the legal system and the administrative regulatory process for a better understanding of the environment in which disputes, affecting the engineer's professional and business activities, are resolved and policy is formulated. The growing thrust of strict liability, the codification of commercial law, the extensive development of social legislation with respect to employee safety, and the rush of environmental regulation illustrate the daily contact engineers have with the legal system and administrative process. In addition, there is growing evidence of the need to give formal consideration to the ethical obligations of professional engineers by noting the existing relationship between ethical and legal concepts.

This book, which is addressed to technically trained engineering students and other students with reasonable sophistication but without substantial legal background, is meant to be studied with reflective analysis and not merely be read. Experience indicates that the case method of presentation with lively give and take discussion is superior to a mere recitation of rules. Actual court and administrative decisions turn on particular facts and procedural rules, and the adversary system is frequently called on to resolve disputes involving highly charged conflicting social, political, or economic interests where both positions represent legitimate claims. This conceptual method, used as the foundation of legal training, serves as a unique bridge to demonstrate that the different areas of the law are interrelated.

As an approach to interdisciplinary teaching, several edited court or administrative decisions are included at the end of each chapter to provide the student with an opportunity to evaluate critically the court's reasoning and also to synthesize a series of cases leading to an identification of policy trends. Because studying edited cases may be a new pedagogical experience for many using this book, the reasons underlying this method should be clearly understood to include (1) a first-hand exposure to legal reasoning and conceptual views of problem solving and (2) a demonstration by concrete illustrations of the importance of fact development as a part of formulating policy options in the decision process. In those edited cases, string citations and footnotes, which are so routine in much legal writing, have been largely omitted because of the rather instinctive but distracting tendency for eye-flitting from the text to the bottom of the page. However, a number of cases are summarized in the body of the text material, and sufficient citations are presented to provide the curious and industrious with additional avenues of study to satisfy the intellectual appetite.

This book is not an attempt to make engineers into lawyers or to be used as a do-it-yourself legal guidebook. Rather, the broader and more philosophical purpose is (1) to facilitate the interrelationship between engineers and lawyers; (2) to indicate the numerous economic, technological, social, and political interests that must be balanced in the judicial resolution of disputes; and (3) to encourage effective participation by engineers in the ongoing regulatory process by which the trend of engineering jurisprudence emerges. This book seeks an approach to the common problem resulting from our extremely specialized education and research by providing a broader perspective balancing that highly narrow technical concentration with the complex and interrelated aspects of our society.

The questions at the end of each chapter serve as a student review, to suggest areas of discussion to be pursued by the instructor, and to raise policy-oriented issues.

The initial phase of this book started while the author was on leave as Director of the Energy Studies Program, Law Center, at the University of Houston, serving as a visiting professor in the School of Engineering at the University of California (Irvine). This opportunity to teach and work with students of another discipline provided a long-awaited opportunity to experiment in the field referred frequently to as interdisciplinary but for which the author prefers to use the term "counterdisciplinary" for the reasons explained in chapter 1.

Finally, this book was not written in a vacuum. My wife, Ellen, was supportive in my research and served as a valued critic of both my writing and my ideas. Frequently, she reminded me that my intended audience were not law students but engineering students. Such criticism and encouragement were important elements in the evolution of these materials.

ACKNOWLEDGMENTS

The excerpts from judicial decisions in the text and the edited opinions appearing at the end of the chapters, are reprinted with the permission of West Publishing Company, St. Paul, Minnesota, the publisher of the National Reporter System.

1

ENGINEERING PRACTICE AND PROFESSIONALISM
Legal and Ethical Issues

1.1 WHY LAW FOR ENGINEERS?

It may seem unusual that engineers should have an interest in legal matters or a concern with understanding the operation of the administrative process or the judicial system. However, knowledge of the basic structure and process of the legal system is not only desirable but also necessary for an efficient and intelligent practice of the engineering profession today. These materials are designed to provide a conceptual overview of the judicial process and the more significant legal principles, policies, and trends that affect the professional engineer.

Interaction of Law with Professional Engineers

Individuals employed or engaged in business are necessarily affected by a whole range of legal rules and regulations. Engineers, whether serving in a professional, technical capacity or as a part of management in a business enterprise, are concerned about compliance with laws, administrative regulations, and court decisions; those negotiating contracts or engaging in professional activities must understand the legal concepts of contractual rights and obligations. Likewise, awareness of the degree of care required in the performance of professional engineering activities is an essential element of prudent job performance. The importance of law in business and for the professional engineer is demonstrated by the marked increase in the use of lawyers by business. Consequently, corporate and business law is the fastest growing segment of legal practice.

For many years the engineering profession was relatively shielded from ethical questions and legal challenges. By voluntary action, various professional engineering societies initiated ethical codes covering standards of individual conduct and in some cases established criteria for design and performance. For example, the American Society of Mechanical Engineers (ASME) developed a Boiler Code, resulting in an outstanding record of safety. Through successful performance, based on conduct dictated by standards similar to the Boiler Code as well as numerous remarkable technological developments produced by the engineering profession, a public perception emerged that engineering technology was almost infallible. However, a number of widely publicized accidents, coupled with the growing tide of consumerism, tarnished this perception, and in its place there developed a wave of legislation, regulations, and decisional law affecting engineering work and products.

Public Perception of Professional Engineers

Examples of engineering miscalculations, while small in percentage, are dramatic in the eyes of the consuming public. The widespread public concern was heightened by such events as the failure of NASA's space shuttle, Challenger; the malfunction at the Three Mile Island nuclear plant, the collapse of the cross walk at a Kansas City hotel, the collapse of the Tacoma Narrows bridge, the problems inherent in the numerous recalls of automobiles, and the almost daily incidents of toxic spills and problems with hazardous waste disposal.

Given the profit motive of manufacturing and the competitive nature of the American free enterprise system, new legal, social, and political problems now confront the engineering community in the day-to-day performance of their activities. Challenging the decision of other professionals, including management decisions, raises new difficult ethical and murky legal dilemmas. This so called whistle-blowing by individuals results in painful introspection and sometimes ultimately blackens professional and corporate images.

In the last quarter of a century, there has been a material change in the public attitude toward those engaged in business and professional activities. Accepted doctrines of *laissez faire* (non interference) and *caveat emptor* (let the buyer beware) have given way to a new wave of consumer protectionism and regulatory legislation. There has been a virtual explosion of product liability litigation, and, today, it is virtually impossible to obtain adequate public liability insurance protection.

Without attempting to evaluate the reasonableness of today's consumer demands, a look at recent statistics of product performance is enlightening. Some 20 million Americans are injured each year from incidents connected with products. Of this number, approximately 30,000 are killed and 100,000 are permanently disabled. These statistics are shocking, even though a number of these incidents may have been caused by improper use of the product or carelessness of the user.

1.2 CONCEPT OF PREVENTIVE LAW

Formerly, lawyers were associated principally with collection of business debts, with lawsuits for nonperformance of contracts, and for damages resulting from personal injury. However, an increasing number of business managers and professionals retain lawyers or seek their advice to help plan business ventures to avoid delays, shutdowns, and penalties through orderly and timely compliance with the rapidly growing volume of legal rules and regulations imposed on business and professional activities. This development has been termed "preventive law."

The term, "preventive law," was coined by Louis M. Brown, professor of law at the University of Southern California, in his book, *Preventive Law,* published in 1950 by Prentice Hall. The goals of preventive law are defined as an arrangement of business plans and methods to increase profits by avoiding losses through fines and damage judgments and by reaching business objectives through enforceable contracts while avoiding governmental prohibitions. Preventing legal pitfalls entails seeking advice in advance rather than after disputes arise. The aims are to reach desired objectives with minimal delay and risk.

Broadened Responsibilities of Engineers

All business and professional activities involve a degree of legal risk, since such activities entail respective rights and duties. Those rights and duties give the parties expectations and obligations that may be enforced within the judicial system. Thus, a broadened perspective opens up for the engineer. The scope of the engineer's responsibilities includes not only the traditional issues of technology but now are coupled also with societal expectations concerning project performance and safety. Since our jurisprudence reflects societal expectations, engineering practice involves basic principles of law and ethics.

The technologies of nuclear energy, computers, space exploration, biochemistry, and environmental concerns have led to a confrontation of the engineering community with the public welfare and with various social issues such as energy conservation, environmental concerns, and product safety. The horizons of engineering are no longer localized but have widened, and new technologies pose problems in which potential resulting errors not only may last for years and be irreversible but may have regional, national, or global implications.

Trend from *Caveat Emptor* to *Caveat Venditor*

Just as traditional interaction between society and business has changed from the spirit of *caveat emptor*, the atmosphere of engineering practice and engineering education has changed. Formerly, the priorities involved strictly technical considerations such as strength of materials, electrical components, manufacturability, marketability, and cost; but, now the wave of consumerism has invoked the spirit of *caveat venditor* ("let the seller beware"). This trend developed a far broader new criteria that include failure analysis, product safety, environmental audits, hazard assessments, and risk evaluation and compliance with numerous industry and governmental standards. All of these areas have a legal regulatory content.

1.3   GOALS OF INTERDISCIPLINARY LEGAL STUDIES

Purposes of Multidisciplinary Communication

It is not the intention of these materials that the reader will learn the law, nor will they prepare the reader to be a lawyer. The mastery of the legal profession requires years of concentrated and specialized study. Sound legal advice requires a competent lawyer. At most, these materials will provide only a conceptual overview. On the other hand, their study may aid in ascertaining the need for legal counsel and assist in using, more effectively, such legal advice. Today's society is so complex that an understanding of other disciplines and communicating effectively across disciplinary borders is essential. The client needs to know what information is relevant or critical to the lawyer's opinion. A legal opinion based on inaccurate facts or incomplete understanding of the activities involved is surely to be of questionable value.

The first aim is to enhance the usefulness of the legal advice. In turn, improvement in the efficiency of the professional engineering practice will follow.

There is a second broader aim. Law represents an important aspect of the culture of our society. It reflects historical and policy solutions to current societal problems. Dealing with law as a developing institution will aid in understanding the social, political, and economic expectations in which business and professional activities are con-ducted. Frequently, disputes arise where both parties have legitimate claims, and these must be resolved by a balancing of conflicting inter-ests.

Many who read these materials will not have a legal education but will have training in science, engineering, or business. These readers should remember that scientists and lawyers work with different bodies of laws. Laws of nature, such as the law of gravity, which the scientist discovers and demonstrates, are inviolable. On the other hand, laws that result from legislation and court decisions do not have the same characteristics. Legal obligations flowing from laws may be broken or ignored and imposition of any sanctions depends on the authority of the judiciary for remedy.

Counterdisciplinary Approach: Technique of Problem Perspective

Frequently, so-called interdisciplinary courses fail because they are either too easy or are overly difficult. In one case, the subject is treated superficially or is so compacted that the student can do little more than memorize a series of rules. In the other case, the materials are cast in such a shroud of legalism as to be incomprehensible except to an advanced law student. This particular set of materials is designed to help students, who have acquired a certain knowledge of one dis-cipline, integrate that knowledge with the unfamiliar discipline in a few hours of a single course despite the students' sophistication in one area and lack of familiarity with the other.

These materials largely ignore the home discipline and con-centrate on the unfamiliar one by emphasizing its process and concep-tual viewpoint. This pedagogy has been termed a "counterdisciplinary" approach. To obtain the maximum impact, actual legal decisions (court and administrative cases) are chosen in settings and fact situations familiar to the engineering profession. This technique emphasizes the way important issues in the engineering field are perceived in the legal arena, the discipline where such disputes are resolved. This technique is considered preferable to the typical interdisciplinary practice in which the law-engineering course becomes a specialized course in con-tracts and torts for the engineer, imparting little information as to the functions of administrative agencies and courts and even less about the jurisprudential process. In dealing with the counterdisciplinary ter-minology and techniques, the primary emphasis is on how the problem was perceived by the new discipline, how the problem was approached, and how it was resolved. To this end, these materials provide an under-standing of the sources of law and distinguish between legal and ethical obligations. Study of the administrative process and judicial system demonstrates the relationships between administrative and judicial officials, the dichotomy between state and federal jurisdictional matters, and the respective functions of trial and appellate courts. In addition, the materials will help explain some of the obscure reasons for seemingly strange results in certain litigated cases.

The reader should be aware of the old axiom that "a little knowledge is dangerous." However, this introduction to legal principles, and an understanding of the way another discipline (i.e., the legal profession) perceives the legal aspects of engineering questions, will enable the engineer more effectively to seek and use legal advice. To this end, some of the materials are based on actual case studies following the traditional method of law study, and these demonstrate that merely memorizing rules is not the ultimate objective. Exceptions are born of generalities. The goal is the ability to distinguish and relate broad generalities to actual fact situations and to synthesize the underlying rationale of decisions to predict trends and probable results in new situations. Generalities are difficult to sustain because the specific factual situation determines the actual application and, therefore, the results.

Goals of Counterdisciplinary Approach

The counterdisciplinary approach eliminates the concept of learning certain "black letter" principles of contract and negligence law and focuses on trends of judicial thinking and policy implications of the legal system within broader societal interaction. The purpose can be summarized as giving the reader some substantive law but focusing more on what law is, reflecting changes of policy, and how law works. Perhaps the most important result will not be the avoidance of damage claims for negligence or contract breach but the interaction, as a member of the engineering profession, in helping move the law and regulatory process in a desirable direction with respect to the reader's profession. A prospectus for a course entitled "Law Course for Medical Students" by Professor Peter Linzer is a relevant and thoughtful statement:

> The purpose of the course is to give [engineering] students some understanding of law and the legal process. It will concentrate on aspects of law most likely to affect the [engineering] practitioner; while students will learn some substantive law, the instructor is much more interested in their understanding what law is and how it works than he is in their learning rules on how to avoid malpractice claims. The latter (primarily: "Be careful") will change in detail many times during the students'

careers; but if the students understand the process, they will be able to deal with changes in specific rules of law, and perhaps more important, will be able to interact with lawyers and to help shape the law as it affects [engineering]. 55 *Journal of Urban Law*, 113, 115 (1977).

1.4 DEFINITIONS OF LAW AND ITS FUNCTIONS

Jurisprudential scholars do not agree on a single definition of law. Some laypeople think of the law on the basis of their experience with the police, while others think in terms of a court trial, damages for negligent conduct, or a real estate transaction.

Law: Basic Conceptual Definitions

Over the centuries, philosophers have discussed the nature of law and the related concept of justice. Four different but commonly accepted basic concepts of law follow:

1. *Law as what is right.* Under this concept, there is some accepted code of what is right and wrong. The source may be divine or human authority. This notion of law bears the closest relation to general understanding of ethics.
2. *Law as custom.* Under this concept, law is a set of rules resulting from accumulated customs and traditions.
3. *Law as command.* Under this concept, law is a body of rules issued by a political authority and enforced by sanctions.
4. *Law as social engineering.* Under this concept, law is a means of social control that balances conflicting interests and values for the common good. Implied in this definition is the dynamic role of law as an instrument of social, political, and economic change. This notion of law represents the greatest distance from the commonly held view of ethics, since a law involving economic regulation seeking to deal with the distribution of economic power may be neutral as to right or wrong.

Lay people conceive of the law in different ways. Most commonly, many consider the law as a set of rules. Thus, the law might be defined as a set of principles or standards of conduct that have general application in the society, that have been developed properly by a competent authority within that society, and that for the violation of which, society imposes a penalty or provides a remedy. On the other hand, rather than dealing with the law in such a broad philosophical sense, it may be considered in a very narrow sense, so that law may refer to a specific branch of law such as product liability or specific statutes such as the Clean Air Act.

Underlying Philosophy of Jurisprudence

The underlying philosophy of jurisprudence is to establish standards of conduct and maintain order. It may serve to redress injustice by reordering the balance of power between different social or economic groups such as employers and workers under the National Labor Relations Act. For the business community, law permits planning by ensuring a degree of certainty with respect to expectations and responses.

1.5 OVERVIEW OF JUDICIAL FUNCTION

Within the U.S. legal system, law is enacted and enforced by government. In addition, the law itself defines and establishes the framework of the government. The underpinning of this system is a doctrine of constitutional law. Fearing the awesome power of government, the framers of the U.S. Constitution sought to safeguard individual rights and freedom from arbitrary governmental action by establishing a system of checks and balances by means of three equal but separate branches of government. This system of checks and balances is a fundamental principle in constitutional law as it was institutionalized among the branches of the federal government. For example, the power of the legislature to enact laws is subject to the review of the judiciary concerning the constitutionality of those laws. To avoid ill-advised legislation, the Constitution requires that the Congress and the president agree on the adoption of new laws. The president is given veto power that can be overridden only with a two-thirds majority in each house of the Congress. The executive is given authority to administer the laws, but their enforcement generally takes the concurrence of the judiciary. In addition, the U.S. Constitution granted certain powers to the federal government but retained certain powers in the states and the people. The Tenth Amendment provides

> The powers not delegated to the United States by the Constitution, nor prohibited by it to the States, are reserved to the States respectively, or to the people.

In addition to the powers and restrictions between the federal and state governments and the people, each state has a constitutional form of government providing for the law of that state being the supreme law of that jurisdiction.

A unique feature of the U.S. legal system is that there are really 51 separate legal systems. This is the result of the federal government with its own system and then each state with its own system. Each jurisdiction exercises sovereign authority to legislate, regulate, and enforce legal principles within its respective jurisdiction. For example, control over interstate and foreign commerce is vested in the federal government, and this is the basis of authority for much federal regulation. Under that power, the federal government regulates the sale of securities under the Securities Act, collective bargaining under the National Labor Relations Act, and pollution control under the Clean Air Act and other environmental laws. State laws are applicable in the sphere of certain local matters such as the ownership of real estate, zoning and building permits, damages resulting from negligence, and breach of contracts. Not infrequently, cases arise posing questions as to which sovereign body (i.e., federal or state) has authority to regulate in a particular field or area.

In *Bethlehem Steel Corp. v. Board of Commissioners,* 276 Cal. App.2d 221, 80 Cal. Rptr. 800 (1969), the court invalidated the California Buy American Act on the grounds that the statute invaded the exclusive power of the federal government. The court conceded that the motive to provide legislative approval of protectionist measures for locally manufactured products was laudable and that California had authority to chose its own procurement policies but held the state could not invade the exclusive jurisdiction of the federal government to conduct "foreign policy."

1.6 SEPARATION OF POWERS DOCTRINE: JUDICIAL FUNCTION

A basic concept of the U.S. Constitution and the various state constitutions is the separation of powers among the legislative, executive, and judicial branches of government. This principle of government affects the legal system in a number of ways. The judiciary does not legislate and does not perform the functions of the executive. Thus, not every dispute is of the nature that the judiciary will entertain or resolve.

Within the federal court system, this principle flows from Section 2 of Article III of the U.S. Constitution that refers to the authority of the courts to hear and decide "controversies." This has been interpreted by the court as prohibiting the judiciary from deciding political questions or giving advisory opinions.

The law that applies to certain legal questions such as negligent conduct or breach of contract may be the result of a combination of laws made by the legislature and judges. The kind of law that results from legislative action is known as "statutory." Statutes are passed by elected representatives after a bill has been introduced into the U.S. Congress or a state legislature. Every year numerous laws are enacted, many of which affect particular industries. As these bills are presented and considered by the legislative body, hearings may be held in which interest groups may present views in support of or in opposition to the proposed legislation. For example, a mechanics lien law may be of interest to contractors, labor unions, owner-developers, and financing institutions. The elected members of the legislature may permit interested parties to testify concerning the proposed legislation, but there is no constitutional or generally any other legal right to appear before the legislature.

While legislation is the function of the legislature and the judiciary is limited to ascertaining the validity of, to construing, and to enforcing the law, legislation does not provide solutions to all issues raised by litigants. It is within this area that the judiciary through the judge must apply principles including those set down by other judges in earlier cases solving similar problems. This practice of following earlier decisions is referred to by the term, *stare decisis*, which means "let the decision stand." This doctrine tends to establish predictability or a "rule of law." To the scientist, there is an analogy to the way the laws of nature operate, reasoning that like causes produce predictable like results.

To illustrate the separation of powers doctrine, the Supreme Court's decision in *Bowsher v. Synar* struck down part of the Gramm-Rudman deficit reduction law as unconstitutional because the comptroller general, a legislative official subject to removal by Congress, had been unconstitutionally assigned executive powers in the deficit-reduction process. The Gramm-Rudman law set deficit ceilings for fiscal years 1986 through 1991, and if the projected deficit for any fiscal year exceeded the statutory maximum, the law required across-the-board cuts in most federal programs to reach the target deficit. The crux of the lawsuit was the comptroller general's role in projecting the target deficit for each fiscal year. The comptroller general, appointed by the president from congressional nominees, could be removed from office only by impeachment or joint resolution of Congress based on five statutory criteria, including neglect of duty, inefficiency, or malfeasance. Under Gramm-Rudman, the comptroller general is charged with preparing a final report for the president on the needed cuts, and this law required that the report be the basis for a presidential sequestration order imposing those cuts. After a fixed period in which

Congress could reduce the spending to meet the deficit target, the sequestration order would take effect. A majority of the Court found that the comptroller general was "an officer, responsible to [Congress] alone . . . [and] . . . to permit an officer controlled by Congress to execute the laws would be, in essence, to permit a congressional veto . . . [since] Congress could remove or threaten to remove an officer for executing the laws in a fashion found unsatisfactory to Congress." Justice Byron White, dissenting, urged the Court to limit its role in separation of powers disputes "to determining whether the Act so alters the balance of authority among the branches of government as to pose a genuine threat to the basic division between the lawmaking power and the power to execute the law." In reaching its decision and in applying the doctrine of *stare decisis*, the Court relied on its earlier decision in *Northern Pipeline Construction Company v. Marathon Pipeline Company*, 458 U.S. 50 (1982), holding that Congress could not give general jurisdictional authority to bankruptcy judges without life tenure, and in *INS v. Chadha*, 462 U.S. 919 (1983), holding invalid the legislative veto of executive action.

1.7 WHAT CONSTITUTES PROFESSIONAL STATUS?

Engineers are subject to a variety of legal obligations and, under the requirements of their practice, may need to meet registration or license requirements and comply with a code of ethics. Although some form of engineering existed as early as the ancient pyramids and the aqueducts of Rome, engineering has been recognized as a separate calling for about a century and a half. In 1750, John Smeaton, an English engineer, made the first recorded use of the term "civil engineering." The Institution of Civil Engineers was founded in Great Britain in 1818.

Defining a Professional

What makes a calling a profession, and what is the significance of that status? For the purposes of labor law, the Taft-Hartley Act ignores the public aspect generally considered a pivotal factor by defining a professional employee, primarily for determining bargaining status, as

> any employee, engaged in work (1) predominantly intellectual and varied in character as opposed to routine mental, manual, mechanical, or physical work; (2) involving consistent exercise of discretion and judgment in its performance; (3) of such a character that the output produced or the

result accomplished cannot be standardized in relation to a given period of time; (4) requiring knowledge of an advanced type in a field of science or learning customarily acquired by a prolonged course of specialized intellectual instruction and study in an institution of higher learning . . . as distinguished from a general academic education or from an apprenticeship or from training in the performance of routine mental, manual or physical processes . . . .

Further refinement by the courts suggests a person is not a professional by merely graduating from a school or college and becoming a member of the society representing that learning or calling. Traditionally, courts in defining a "professional" look to a variety of factors including registration requirements for practicing the calling, representation of members and control of activities by a formally organized society, adherence to a code of ethics, and the public service nature of the occupation. Generally, the critical test is based on the obligation to the public.

Professional Codes and Ethics

The Engineers Council for Professional Development (ECPD), which represents 12 major engineering societies, adopted a Code of Ethics of Engineers. These canons of ethics focus on the engineer's relationship with other engineers, employers, and the public. It is this latter constituency that is critical to the professional standing and is a fundamental principle articulated as ". . . being honest and impartial, and serving with fidelity the public . . . ." The first of seven fundamental canons is stated as "Engineers shall hold paramount the safety, health and welfare of the public in the performance of their professional duties."

Ethical considerations are raised in a variety of settings of which product safety frequently involves the balancing of a safe product that must be sold at a price unacceptable to the consumer, in turn driving the customers to a less expensive and less safe product, versus a product that maximizes the monetary rate of return although having certain inherent faults considered as a calculable risk assessment. Such situations place competing interests and loyalties in conflict. One duty is to the employer and the other is to the public.

Modern society is built upon an enterprise organizational form to conduct its business. Typically, almost the entire range of activities such as research, designing, construction, manufacturing, and marketing are organized into large industrial units with multiple layers of supervision and where various subdivisions of professional managers are responsible for ultimate operating decisions. Since this bureaucracy is the dominant institutional form of modern industrial society, virtually every occupation and profession is specialized, certified, and arranged into a hierarchy and a bureaucracy. Under this structured organization form, the free exercise of professional fidelity to the public may conflict with explicit job descriptions or defined authority. This presents a challenging dilemma to the conscientious professional where the job description limits the decision responsibility to a narrow range of issues.

Whistle-Blowing

There have been a number of well publicized cases in which engineers have publicly accused their employers of proceeding with development of unsafe products or projects. These problems are never easy or simple since they involve a variety of fact situations and complex judgmental determinations. It is possible that, in the larger context, individual engineers may not have all the facts bearing on the decisional process. Supervisory employees or upper-level management may arrive at a different opinion since their usual function is to select options from the range of alternatives none of which may be perfect. Nevertheless, the expression of a differing professional view is appropriate if held and offered in good faith and based on credible data and information. Within routine conduct, the expression of a dissenting opinion is proper professional conduct. Public dissent, the expression of a differing view and popularized under the term "whistle-blowing," is generally defined as the disclosure to one in authority of a violation of law or threat to public health and safety. By definition, whistle-blowing is an extraordinary act, usually requiring an employee to go outside of the established organizational channels with a concern related to the public interests.

The legal status of the whistle-blowing employee has been in considerable flux based both on trends in statutory protection and changes in case law. Since the late 19th century, employers have been able to dismiss without liability "employees-at-will" (an employee with no definite term of employment). Further, the employer was not required to give any reason for the termination. Underlying the employment-at-will doctrine was the legal concept to maximize freedom of contract and the notion that both parties benefited from being able (as the agreement appeared to intend) to terminate the relationship freely. Employees were able to move to a better position and the employer had flexibility to respond to business needs.

Starting in the 1930s, Congress implemented other statutory policies such as the right to unionize, by protecting employees who were discriminated against or terminated because of complaining or testifying against the employer in connection with a charge of violation of the labor laws. Similar provisions have been incorporated in other federal and state laws. In the 1960s, both federal and state laws moved to protect group rights by precluding employers from discriminating on the basis of race, color, sex, nationality, religion, age, and, to some extent, handicap. Section 11(c) of the Occupational Safety and Health Act (OSHA) prohibits an employer from discharging or discriminating against an employee who exercises any rights under OSHA, such as employees complaining to OSHA about what they believe to be hazardous conditions on the worksite. The False Claims Act (1968) authorizes filing private lawsuits for alleged fraudulent practices by government contractors permitting a recovery of three times the fraudulent overcharge and allowing an award to the whistle-blower of 15 percent of the amount recovered by the government plus attorney's fees. This statutory trend was accompanied with a similar movement in the case law that created exceptions to the employment-at-will doctrine where a particular discharge violated a significant and identifiable public policy.

Traditionally, the law sided with management when dissent spilled over into (perceived) insubordination. But, under a developing trend, the law has increasingly intervened in defense of responsible employee dissent in situations where the employee acts to further an acknowledged public interest. Thus, the courts began to create limited exceptions to the dismissal-at-will doctrine, where a particular discharge or discipline violated a significant and identifiable public policy. For example, damages were awarded to employees who were dismissed for performing jury duty, *Reuther v. Fowler & Williams, Inc.*, 386 A.2d 119 (Pa., 1978); for filing workers' compensation claims, *Frampton v. Central Indiana Gas Co.*, 297 N.E.2d 425 (Ind. 1973); urging employer compliance with the Food and Drug Act, *Sheets v. Teddy's Frosted Foods, Inc.*, 427 A.2d 385 (Conn. 1980); reporting overcharges on accounts to bank auditors, *Harless v. First National Bank in Fairmont*, 246 S.E.2d 270 (W. Va. 1978); and reporting shortcomings at a nuclear plant by a quality control inspector, *Atchison v. Brown & Root, Inc.*, 82 ERA 9.

1.8 PROFESSIONAL ETHICS

Ethics is a branch of philosophy; it is moral philosophy or thought concerning moral problems and moral judgments. This general rule may be articulated in such conduct as we ought never to harm anyone, we ought to keep our promises, and we ought to obey authority. The terms "moral" and "ethics" frequently are used interchangeably and generally are equated to "right" and "good" as opposed to "wrong" and "immoral."

Ethics refers to imperatives regarding the welfare of others in a more general and impersonal sense than morals. In his book, *Moral Philosophy*, published in 1964 by Charles Scribner' Sons, New York City, J. Martitain presents a historical and critical survey of the great systems of moral philosophy and notes that ethics coordinates with but is different from religion. However, many attribute a religious biblical underpinning to ethics and point to the commandment of the Golden Rule. There is also reported a similar precept in Confucian literature where in response to a request from a disciple for a single word that could serve as a guide to conduct for one's life, Confucius is recorded as replying, "Is not reciprocity such a word?"

Great philosophers have argued the egoistic or utilitarian theories in respect to such questions of breaking promises and truthfulness concerning the universality of rules versus exceptions to rules. But the common thread through such arguments is that ethics, in one sense, is a social enterprise, not discovered or invented (e.g., adopted as a law) by the individual for guidance but existing for the individual as language. It is social in its origins, sanctions, and functions. It is an instrument of society as a whole for the guidance of individuals and groups.

Ethics is to be distinguished from "prudence," although both may dictate some of the same conduct, just as law may frequently require the same action. Ethical conduct is a function of Freud's super-ego, which does not think merely in terms of getting what is desired by the individual *id* or even in terms of salvaging the greatest balance of satisfaction over frustration for it.

An "ethical professional or businessperson" has been defined "as a person who is the full measure of what a person should be" (P. Heyne, *Private Keepers of the Public Interest*, McGraw-Hill, New York, 1968, p. 113). Difficult issues are raised for the professional engineer facing the dilemma of conflicts in professional judgment and loyalties to the family, employer, fellow workers, and the public. What are the responsibilities of engineers to their employers, to their clients, to those who use or consume the products and services they provide? When is it wrong to just do the job and leave the decisions to management? What is the responsibility of professional engineers to blow the whistle when they know that something is not being done properly?

Former Federal Reserve chairman Paul Volcker in a discussion of business ethics reported in the *Houston Post*, Sept. 21, 1986, on page E9, stated, "I do have a feeling that people are willing to sail out closer to the wind these days than they were 20-30 years ago . . . ." This leads to the question of whether cutting ethical corners is now deemed acceptable policy under the test that where one ends up matters far more than how one got there.

Professionals may be subject to certain legal registration requirements that seek to assure the public of the applicant's training and
competency, but professional societies also may set out higher and more
specific rules of conduct in professional practice. These codes or canons
of ethics may dictate more exacting criteria than legislative laws or
court decisions.

Carl F. Taeusch, in *Professional and Business Ethics*, published in
1926 by Holt and Co., New York City, describes professional ethics ". . .
as a set of principles of professional conduct which have evolved by
practice and by intellectual analysis for the guidance of the members of
the profession for the greatest good of the profession and the public at
large." He concludes that in any question of professional ethics, three
major elements are to be considered in the following order: (1) the effect on the welfare of the public, (2) the effect on the other members of
the profession and the profession as a whole, and (3) giving adequate
consideration to the matter of loyalty to the employer.

Recurring newspaper and magazine articles refer to the breach of
professional ethics that affect the welfare of the public. The disclosure
in 1986 of the widespread improper use of inside information by
various stockbrokers, bankers, and attorneys had a devasting effect on
the public perception of the securities markets. The so-called Levine insider trading case led to criminal allegations against a number of well
known figures in the financial field. A partner of a prominent New
York law firm entered a guilty plea to charges of passing to an insider
trading ring information concerning pending takeover deals acquired in
his legal practice. Such conduct has resulted in a loss of public trust in
fundamental business institutions as reflected in a 1985 study, which
found little or nothing to ensure ethical behavior and concluded that
the message was "if the bottom line is right, go for it." (Harry Stein,
"The Struggle Not to Sell Out," *Esquire*, Sept. 1985, p. 35). On the
other hand the problem has been a matter of concern for some time, as
shown by the observations in 1934 by Ivan Lee Holt, a Methodist
bishop, writing in *The Return of Spring to Man's Soul*, published by
Harper & Brothers that, "We take our ethics as we take our athletics
— watching [others] play the game, to be commended now, and to be
criticised later where there is a failure to give the best or abide by the
rules. A sense of obligation has slipped from our shoulders. We have
wanted the fruits of civilization, but none of its burdens. Downright
honesty we must have, or else our whole social structure collapses."

For a number of years, courses dealing with professional responsibilities and legal ethics have been an accepted part of the law school
curriculum. The wide-spread disclosures of not only questionable ethical but also criminal conduct by corporate executives has given impetus
to modify the curriculum of business schools to add courses with ethical

content. Rapid advances in medical science coupled with staggering increases in the cost of providing adequate health services has caused the medical profession to reexamine and give new emphasis to medical ethics. This trend appears destined to invade and is invading academic circles in a number of fields. It is only natural that the curriculum of engineering schools should be reexamined and influenced by this movement.

Observing the moral obligation to the professional institution requires high sensitivity to ethical concerns that may involve supporting unpopular positions and ignoring short-term criticism. Peer scrutiny and regulation have been developed to a high degree in the legal and medical profession. However, examples of abuses and problems of implementation are recorded in many instances. An example involving the medical profession, where peer review serves as the principal tool for regulating doctors with respect to professional competency, illustrates some issues raised by peer review. In that case a surgeon, X, after receiving an unsatisfactory evaluation by a peer review panel, brought an anti-trust suit against the peer group alleging inattention by the group to his defense presentation. X charged that the doctors on the peer panel were trying to force the plaintiff out of business since he was competing with them for patients.

Indeed, following the award of anti-trust damages against the members of the peer group in this case, many fellow physicians refused to participate in subsequent peer reviews and became mute since anti-trust cases are hard to defend and the damages awarded in such cases are not covered by the doctor's malpractice insurance. Effective self-policing and professional regulation requires that the professional must overcome the reluctance to report, investigate, present testimony against, and discipline their colleagues who engage in wrong doing or who do not have the required level of competency.

While loyalty to one's master historically dates back to biblical times, the examples of conflict of interest and double dealing are commonplace in today's press. Examples include an article reporting that Congressional representatives failed to disclose fully consulting fees from a developer and to report honorariums from industries seeking to avoid increased regulation.

Ethical and moral values are principles or standards "by which human actions are judged right or wrong" (1986 *Britannica Book of the Year*, p. 34). These standards may or may not have parallel legal obligations or proscriptions. Traditional values of honesty, integrity, and attention to the rights and needs of others are still the best guidelines for making ethical decisions in the business and professional world. Thomas Huxley, in his essay, "Evolution and Ethics," published in 1970 by ANS Press, New York City, wrote

[T]he practice of that which is ethically best . . . involves a course of conduct, which in all aspects, is opposed to that which leads to success in the cosmic struggle for existence. In place of ruthless self-assertion it demands self-restraint; in place of thrusting aside, or treading down, all competitors, it requires that the individual shall not merely respect but shall help his fellows; its influence is directed, not so much to the survival of the fittest, as to the fitting of as many as possible to survive. It repudiates the gladiator theory of existence. It demands that each man who enters into enjoyment of the advantages of the policy shall be mindful of his debt to those who have laboriously constructed it; and shall take heed that no act of his weakens the fabric in which he is permitted to live. Laws and moral precepts are directed to the end of curbing the cosmic process and reminding the individual of his duty to the community, to the protection and influence of which he owes, if not existence itself, at least the life of something better than a brutal savage.

TARASOFF v. REGENTS OF UNIVERSITY OF CALIFORNIA
118 Cal. Rptr. 129, 529 P.2d 553 (1974)
Rehearing Granted March 12, 1975.

TOBRINER, Justice: On October 27, 1969, Prosenjit Poddar killed Tatiana Tarasoff. Plaintiffs, Tatiana's parents, allege that two months earlier Poddar confided his intention to kill Tatiana to Dr. Lawrence Moore, a psychologist employed by the Cowell Memorial Hospital at the University of California at Berkeley. They allege that on Moore's request, the campus police briefly detained Poddar, but released him when he appeared rational. They further claim that Dr. Harvey Powelson, Moore's superior, then directed that no further action be taken to detain Poddar. No one warned Tatiana of her peril.

Concluding that these facts neither set forth causes of action against the therapists and policemen involved, nor against the Regents of the University of California as their employer,the superior court sustained defendants' demurrers to plaintiffs' second amended complaints without leave to amend. This appeal ensued.

Plaintiffs' complaints predicate liability on defendants' failure to warn plaintiffs of the impending danger. Defendants, in turn, assert that they owed no duty of reasonable care to Tatiana and that they are immune from suit under the California Tort Claims Act of 1963.

We shall explain that defendant therapists, merely because Tatiana herself was not their patient, cannot escape liability for failing to exercise due care to warn the endangered Tatiana or those who reasonably could have been expected to notify her of her peril. When a doctor or a psychotherapist, in the exercise of his professional skill and knowledge, determines, should determine, that a warning is essential to avert danger arising from the medical or psychological condition of his patient, he incurs a legal obligation to give that warning. Primarily, the relationship between defendant therapists and Poddar as their patient imposes the described duty to warn.

We reject defendants' asserted defense of Governmental immunity; no specific statutory provision shields them from liability for failure to warn, and Government Code Section 820.2 does not protect defendants' conduct as an exercise of discretion.

In analyzing this contention, we bear in mind that legal duties are not discoverable facts of nature, but merely conclusory expressions that, in cases of a particular type, liability should be imposed for damage done. As stated in *Dillon v. Legg*, 441 P.2d 910, 916 (Cal. 1968), "The assertion that liability . . . be denied because defendant bears no duty to plaintiff begs the essential question [of] whether the plaintiff's interests are entitled to legal protection against the defendant's conduct. [Duty] is not sacrosanct in itself, but only an expression of the sum total of those considerations of policy which lead the law to say that the particular plaintiff is entitled to protection." (Prosser, *Law of Torts*, 3d ed., 1964 at pp. 332-333). *Rowland v. Christian*, 443 P.2d 561, 564 (Cal. 1968), listed the principal considerations: "the foreseeability of harm to the plaintiff, the degree of certainty that the plaintiff suffered injury, the closeness of the connection between the defendant's conduct and the injury suffered, the moral blame attached to the defendant's conduct; the policy of preventing future harm, the extent of the burden to the defendant and consequences to the community of imposing a duty to exercise care with resulting liability for breach, and the availability, cost, and prevalence of insurance for the risk involved."

Although under the common law, as a general rule, one person owed no duty to control the conduct of another . . . nor to warn those endangered by such conduct (Rest.2d Torts, supra, Sect 314, com. c), the courts have noted exceptions to this rule. In two classes of cases the courts have imposed a duty of care: (1) cases in which the defendant stands in some special relationship to either the person whose conduct needs to be controlled or in a relationship to the foreseeable victim of that conduct (Rest.2d Torts, supra, Sects. 315-320); and (2) cases in which the defendant has engaged, or undertaken to engage, in affirmative action to control the anticipated dangerous conduct or protect the prospective victim. (Rest.2d Torts, supra, Sects. 321-324a.) Both exceptions apply to the facts of this case.

Turning, first, to the special relationships present in this case, we note that a relationship of defendant therapists to either Tatiana or to

Poddar will suffice to establish a duty of care; as explained in section 315 of the Restatement (Second of Torts), a duty of care may arise from either "(a) a special relation . . . between the actor and the third person imposes a duty upon the actor to control the third person's conduct, or (b) a special relation . . . between the actor and the other which gives to the other a right to protection."

Although plaintiff's pleadings assert no special relation between Tatiana and defendant therapists, they establish as between Poddar and defendant therapists the special relation that arises between a patient and his doctor or psychotherapist. Such a relationship may support affirmative duties for the benefit of third persons. (See Fleming & Maximov, *The Patient or His Victim: The Therapist's Dilemma,* (1974) 62 Cal. L. Rev. 1025, 1027-1031). Thus, for example, a hospital must exercise reasonable care to control the behavior of a patient which may endanger other persons. A doctor must also warn a patient if the patient's condition or medication renders certain conduct, such as driving a car, dangerous to others.

Although the California decisions that recognize this duty have involved cases in which the defendant stood in a special relationship both to the victim and to the person whose conduct created the danger, we do not think that the duty should logically be constricted to such situations. Decisions of other jurisdictions hold that the single relationship of a doctor to his patient is sufficient to support the duty to use reasonable care to warn of dangers emanating from the patient's illness. The courts hold that a doctor is liable to persons infected by his patient if he negligently fails to diagnose a contagious disease.

More closely on point, since it involved a dangerous mental patient, is the decision in *Merchants Nat. Bank & Trust Co. of Fargo v. United States,* (D.N.D.1967) 272 F. Supp. 409. The Veterans Administration arranged for a patient to work on a local farm, but did not warn the farmer of the man's background. The farmer consequently permitted the patient to come and go freely during non-working hours; the patient borrowed a car, drove to his wife's residence and killed her. Notwithstanding the lack of any "special relationship" between the Veterans Administration and the wife, the court found the Veterans Administration liable for the wrongful death of the wife.

Defendant therapists advance two policy considerations which, they suggest, justify a refusal to impose a duty upon a psychotherapist to warn third parties of danger arising from the violent intentions of his patient. We explain why, in our view, such considerations do not preclude imposition of the duty in question.

First, defendants point out that although therapy patients often express thoughts of violence, they rarely carry out these ideas. Indeed the open and confidential character of psychotherapeutic dialogue encourages patients to voice thoughts, not as a device to reveal hidden danger, but as part of the process of therapy. Certainly a therapist should not be encouraged routinely to reveal such threats to acquaintances of the patient; such disclosures could seriously disrupt the patient's relationship with his therapist and with the persons threatened. In singling out those few patients whose threats of violence present a serious danger and in weighing against this danger the harm to the patient that might result from the revelation, the psychotherapist renders a decision involving a high order of expertise and judgment.

The judgment of the therapist, however, is no more delicate or demanding than the judgment which doctors and professionals must regularly render under accepted rules of responsibility. A professional person is required only to exercise "that reasonable degree of skill, knowledge, and care ordinarily possessed and exercised by members of [the] profession under similar circumstances." As a specialist, the psychotherapist, whether doctor or psychologist, would also be "held to that standard of learning and skill normally possessed by such specialist in the same or similar locality under the same or similar circumstances." But within that broad range in which professional opinion and judgment may differ respecting the proper course of action, the psychotherapist is free to exercise his own best judgment free from liability; proof, aided by hindsight, that he judged wrongly is insufficient to establish liability.

Second, defendants argue that free and open communication is essential to psychotherapy; that "[u]nless a patient . . . is assured that . . . information [revealed by the patient] can and will be held in utmost confidence, he will be reluctant to make the full disclosure upon which diagnosis and treatment . . . depends." The giving of a warning defendants contend, constitutes a breach of trust which entails the revelation of confidential communications.

We recognize the public interest in supporting effective treatment of mental illness and in protecting the rights of patients to privacy and the consequent public importance of safeguarding the confidential character of psychotherapeutic communication. Against this interest, however, we must weigh the public interest in safety from violent assault. The Legislature has undertaken the difficult task of balancing the countervailing concerns. In Evidence Code section 1024, however, the Legislature created a specific exception to the psychotherapist-patient privilege: "There is no privilege . . . if the psychotherapist has reasonable cause to believe that the patient is in such mental or emotional condition as to be dangerous to himself or to the person or property of another and that disclosure of the communication is necessary to prevent the threatened danger."

The revelation of a communication under the above circumstances is not a breach of trust or a violation of professional ethics; as stated in the Principles of Medical Ethics of the American Medical Association (1957) section 9: "A physician may not reveal the confidences entrusted to him in the course of medical attendance . . . unless he is required to do so by law or unless it becomes necessary in order to protect the welfare of the individual or of the community." We conclude that the public policy favoring protection of the confidential character of patient-psychotherapist communications must yield in instances in which disclosure is essential to avert danger to others. The protective privilege ends where the public peril begins.

We turn to the issue of whether defendants are protected by governmental immunity for having failed to warn Tatiana or those who reasonably could have been expected to notify her of her peril. We focus our analysis on Section 820.2 of the Government Code. That provision declares, with exceptions not applicable here, that "a public employee is not liable for an injury resulting from his act or omission where the act or omission was the result of the exercise of the discretion vested in him,whether or not such discretion [was] abused."

Noting that virtually every public act admits of some element of discretion, we drew the line in *Johnson v. State of California*, 447 P.2d 352 (Cal. 1970), between discretionary policy decisions which enjoy statutory immunity and ministerial administrative acts which do not. We concluded that section 820.2 affords immunity only for "basic policy decisions."

We also observed that if courts did not respect this statutory immunity; they would find themselves "in the unseemly position of determining the propriety of decisions expressly entrusted to a coordinate branch of government." It is therefore necessary, we concluded, to "isolate those areas of quasi-legislative policy-making which are sufficiently sensitive to justify a blanket rule that courts will not entertain a tort action alleging that careless conduct contributed to the governmental decision."

Relying on *Johnson*, we conclude that defendants in the present case are not immune from liability for their failure to warn of Tatiana's peril. *Johnson* held that a parole officer's determination whether to warn an adult couple that the prospective foster child had a background of violence "present[ed] no . . . reasons for immunity" and indeed constituted "a classic case for the imposition of tort liability." Although defendants in *Johnson* argued that the decision whether to inform the foster parents of the child's background required the exercise of considerable judgmental skills; we concluded that the state was not immune from liability for the parole officer's failure to warn because such a decision did not rise to the level of a "basic policy decision."

We also noted in *Johnson* that federal courts have consistently categorized failures to warn of latent dangers as falling outside the scope of discretionary omissions immunized by the Federal Tort Claims Act. See *United Air Lines v. Weiner*, 335 F.2d 379, (9th Cir. 1964), decision to conduct military training flights was discretionary but failure to warn commercial airline was not.

We conclude, therefore, that the defendants' failure to warn Tatiana or those who reasonably could have notified her of her peril does not fall within the absolute protection afforded by Section 820.2 of the Government Code. We emphasize that our conclusion does not raise the specter of therapists employed by government indiscriminately held liable for damages despite their exercise of sound professional judgment. We require of publicly employed therapists, that they use that reasonable degree of skill, knowledge, and conscientiousness ordinarily exercised by members of their profession. The imposition of liability in those rare cases in which a public employee falls short of this standard does not contravene the language or purpose of Government Code Section 820.2.

For the reasons stated, we conclude that plaintiffs can assert the elements essential to a cause of action for breach of a duty to warn. The judgment of the superior court dismissing plaintiffs' action is reversed and the cause remanded for further proceedings consistent with the views expressed herein.

DISCUSSION QUESTIONS

1. An effort to revise the ethics code for an Institute of Certified Financial Planners, by proposing a strict prohibition against accepting any type of gratuity or sales incentives (e.g. trips and other prizes), was rejected. A financial corporation sponsoring a real estate partnership offered a free Hawaiian trip to financial planners who achieved a certain level of sales of those partnership interests. One planner argued against the prohibition taking the position that "It's part of the American system that if you produced above average, certain pursuits of your labor should be rewarded, within reason." Discuss the issues involved in the proposed code of ethics.

2. Why should an engineer be concerned about the legal aspects of engineering work? In what kinds of legal proceedings may an engineer be involved?

3. In the *Tarasoff case*, how does the court reconcile the doctor's ethical problem and how does this resolution of the issue affect the doctor's treatment of future mental patients? What recommendations would you have for the doctor if a similar case arose again, and what should the doctor do with respect to the doctor/patient relationship?

4. In the brief description of the Supreme Court decision on the Gramm-Rudman law referred to earlier in this chapter, discuss the legal position of the majority in holding the law unconstitutional. Does the dissent by Justice White border on "the ends justify the means" argument?

5. Discuss the different conceptual jurisprudential definitions of law and how they would affect trends in the law.

6. In designing a large office building for owner O, engineering firm X, used a certain piling for support. Without notifying firm X, owner consults engineering firm Y, seeking advice as to the whether the pilings are necessary or could be eliminated by a better design. Discuss any problems involved and how firm Y should handle this matter.

7. Can a professional employee (engineer, lawyer, or doctor) employed by a large corporation meet professional ethical obligations and also be a loyal employee? In what type of situations will these questions occur?

8. Guidelines and recommendations prepared by the American Society of Civil Engineers (ASCE) for the purpose of prevention failures and producing quality in construction projects are organized in a manual entitled *Quality in the Construction Project* (1988). That manual, addressing the economic aspects of the construction project, concedes that "at times, expectations of cost and quality may appear in conflict, particularly when the time of completion is important or rate of return is marginal" by reasoning that the requirement for early return on investment may arguably compress both design and construction time resulting in lower questionable quality. Acknowledging the economic aspect of construction projects, the manual postulates that acceptable quality does not require perfection. If it is concluded that perfection is neither economical nor practical, how are these two awesome elements (economics and safety) ethically balanced in designing construction projects?

2

THE JUDICIAL SYSTEM
Overview of Legal Sources
and Concepts

2.1 CLASSIFICATIONS OF LAW

While there are many ways to subdivide law, one method is to distinguish between substantive law and procedural law.

Substantive Law and Procedural Law

Substantive law consists of the doctrines that the courts use in deciding cases and may be broken into three subcategories of (1) duties, (2) privileges and (3) powers.

1. Duties are commands to take or refrain from certain action (e.g., driving within the speed limit or improperly storing hazardous waste).
2. Privileges include rights of a natural person, a corporation or other particular form of business enterprise (e.g., the applicability of constitutional rights of free speech or protection from unwarranted search).
3. Powers include specific capacity to take action (e.g., the authority of the agent to bind the principal or the power of an adult to enter into contracts).

Procedural law is a set of rules under which substantive law is enforced. Procedural rules establish the appropriate court for litigating disputes, the proper documents to be filed, the manner in which the trial is conducted, and how judgments and other orders are enforced.

Criminal and Civil Law

Law can also be classified as criminal law and civil law. Criminal law is concerned with a breach of duties owed to society (e.g., embezzlement, arson, or income tax fraud). In criminal law, society through a public prosecutor brings an action in which the sanction may be punishment by imprisonment or a fine. The second category, civil law involves the breach of duties owed to individuals (e.g., breach of a contract or negligence by failing to exercise an appropriate degree of care). In those situations, the judgment is not to punish but to make the injured party whole through the payment of damages. The term "Civil Law" is used in another sense meaning the legal system of most European and Latin American countries and was historically based on Roman law as opposed to the so called English Common Law.

2.2   DEVELOPMENT OF COMMON LAW, EQUITY, AND UNIFORM CODES

The U.S. legal system generally had its roots in Anglo-Saxon jurisprudence, which was referred to as "Common Law." The body of law, existing prior to the Norman conquest of England (1066), was based on "local law," with each town or shire having its own law. Prior to the Norman Conquest of 1066, the English system for resolving disputes was primarily local. Courts existing in that society were in effect under the control of the nobility, the principal property owners, who had a distinct self interest in settling disputes among serfs, vassals and others beholden to them. While those courts developed a system of law, predominantly local in character as to each town and shire, William the Conqueror took immediate steps to organize and unite the entire country by establishing a King's Council and sending royal judges to hold court throughout England. This resulted in the local law being replaced by a uniform system, known as the Common Law. Under this system, the judges instituted a practice of writing their opinions, stating the general principles of law involved and explaining the reasons for their decision. In subsequent cases, judges confronted with similar facts relied on the earlier opinions of other judges, and this practice became known as *stare decisis* ("let the decision stand"). By about the

13th century, the practice of sending royal judges from London to all parts of the realm had a crystallizing effect and produced a uniformity that forged a body of law that was "common" to all England. Subsequently, Justice Henry Bracton collected various decisions rendered under the King's council, which formed the basis of the Common Law of England.

Shortly after the American Revolution, an American jurist, James Kent, wrote a commentary based on the decided cases and described Common Law as ". . . those principles, usages and rules of action applicable to the government and the security of persons and property, which do not rest for their authority upon any express or positive declaration of the will of the legislature." The Common Law, not resulting from legislative enactment, was not a fixed or an absolute set of rules in the sense of statutory law. Common Law was derived from the application of natural reason, an innate sense of justice, and the dictates of conscience. While the Common Law provided a fundamentally constant form of jurisprudence, the system was flexible and adaptable to modifications necessitated by the growing and changing society. In the Common Law, it was quite possible for a judge to change the law by overruling an earlier decision. While it is true that the law does not always keep pace with the changing society or represent current popular notions, those changes must be left to the legislature, which is the representative body. Changes in the judicial law can result only when another case arises that deals with the same point of law, which gives the court a chance to reconsider the earlier decision. This method of modifying the law is limited in usage and occurs when a new case arises in the judicial system, and, despite a precedent of a prior case, the court decides not to follow the prior decision. The court may reason that (1) the facts of the the the instant case are sufficiently different to be distinguishable, (2) the prior decision was "palpably wrong," or (3) various policy reasons now justify a change in the law of the case.

Legal scholars have classified the role assumed by the judiciary into two philosophical categories: judicial restraint and judicial activism. Judges adhering to the restraint philosophy are reluctant to change the law or decide any matters other than those necessary for the immediate decision. Their position rests on the theory that it is the responsibility of the legislature to amend existing law and that the function of the judiciary is limited to interpretation and enforcement. In contrast, the position of the activist philosophy is that the courts are leaders in social, political, and economic movements. To judges within this school of jurisprudence, the U.S. Constitution should be interpreted not as the framers intended when they wrote it but rather in light of what the framers would have intended in relation to contemporary problems. Illustrative of this latter view is the opinion by Chief Justice

Earl Warren in a school segregation case, *Brown v. Board of Education*, 74 S. Ct. 686 (1954):

> In approaching this problem, we [the court] cannot turn back the clock to 1868 when the Amendment [14th amendment] was adopted, even to 1896 when *Plessy v. Ferguson* [separate but equal doctrine] was written. We must consider public education in the light of its full development and its present place in American life throughout the Nation. Only in this way can we determine if segregation in public schools deprives these plaintiffs of equal protection of the laws. Today, education is perhaps the most important function of state and local governments . . . . [Education] is a principal instrument in awakening the child to cultural values, in preparing him for later professional training, and helping him to adjust normally to his environment.

Legal doctrines are also altered by the legislature when it is dissatisfied with the law in a particular field. Frequently, judicial decisions have muddled a specific area of the law, and it is desirable for the legislature to restate or codify the law in coherent fashion. The Uniform laws, adopted in a number of fields of the law, are illustrated in later chapters by reference to codification of the Uniform Commercial Code (UCC) and by reference to the Restatement of various fields of law. Although it was recognized that industry and business do not conform to arbitrary state boundaries, in a number of fields of law the court decisions and statutory law take diametrically opposed positions arbitrarily determined by different state laws. This was especially troublesome with respect to the law covering many commercial transactions where legal doctrine varied from state to state. While this problem was clearly apparent as early as the turn of the century, it was not until 1945, through the joint efforts of lawyers, judges, and law professors, that meaningful action was taken to focus on the creation of coherent integrated statutes to deal with commercial transactions among the states. The result was the initial adoption of the Uniform Commercial Code by Pennsylvania in 1954. By 1972 the UCC was adopted in all states except Louisiana.

2.3 DEVELOPMENT OF EQUITY

"Equity" developed as a separate branch of law as the result of the harshness of the Common Law and the failure of the Common Law to provide all the rights and remedies required by the complexity of a highly developed society. The Common Law was restricted in giving remedies except for injuries falling within the narrow and strictly observed King's original writs (i.e., trespass, trespass on the case, detinue, and replevin). Each of these writs (cases at law) was precisely limited to defined civil wrongs, and no writ was available for wrongful conduct, which varied from the scope of the specific writ. Generally, three remedies were available at Common Law: money damages, return of real property, and return of personal property. In order that justice might be accomplished where these legal remedies were inadequate, injured parties sought relief from the King in person. Since this appeal was to the King's conscience, he referred the cases to his spiritual advisor, the chancellor, usually a church official. In granting a remedy, the chancellor took note of the ecclesiastical law and the Civil (Roman) Law.

This practice became known as "cases in equity" as opposed to "cases at law." While modern practice has now abolished use of a separate equity court docket and system, the dichotomy did exist for many years in the United States.

Decisions of the court in equity are referred to as "decrees" and frequently involve remedies of "injunction" or "specific performance," which are orders to do or not do a particular act. Rules of equity require the court to determine first whether an adequate remedy at law exists and then consider certain equitable maxims:

1. He who comes into equity must do so with clean hands.

2 He who seeks equity must do equity.

3. Equity aids the vigilant.

4. Equity will not suffer a right to exist without a remedy.

For example, an owner hires a contractor to construct a building, and the contractor's excavation near X's adjacent property line threatens to cause cracks in X's building. At Common Law, X can only sue for money damages to remedy the loss. But, in equity, a temporary injunction or restraining order may be issued to stop construction until the nature of the construction is changed to prevent cracking or other irreparable damage.

2.4 SOURCES OF LAW

There are a variety of sources of law in each of the 51 legal systems of the United States (federal and 50 states). Generally, the sources are similar in each of the systems.

Constitutions

The federal government and each of the states have constitutions. While the constitutions will vary in form since some are stated in broad principles and others contain much detail, all represent the basic authority of the sovereign government and are the general charters for governmental authority within that geographic area. All constitutions were originally drafted by a legislative body, usually called a constitutional convention. Drafting, enactment, and subsequent amendments are fundamental examples of the democratic process.

The federal constitution is a grant of power by the people and the states to the federal government, while state constitutions are basically limitations on the power of the states. This conceptual understanding as to the division and limitation of power has given rise to many legal controversies concerning the restrictions on governmental authority and the reservation of power and rights by the people. Freedom of speech is an example of a limitation on sovereign authority and a right reserved to the people.

Treaties

The federal government enters into treaties with other sovereign foreign governments, and the U.S. Constitution provides that treaties made by the president and ratified by the Senate are the "supreme law of the land." Treaties involve subject matter such as recognition of patent rights, reciprocal treatment of foreign taxes, and international trade.

Statutes

Within the federal government and each state, the product of Congress or the state legislature is statutory law. The action by which the legislature takes this action is variously referred to as enacting, adopting, or passing laws. Laws are intended as a general solution to problems determined to exist by the legislature. Thus, statutes establish rules for a wide range of personal conduct and business activity such as the use of automobiles, the sale of land, the licensing of professions, and the prevention of hazardous waste disposal. Frequently, the courts in interpreting a statute will seek to ascertain the legislative intent of a particular law by reviewing the legislative history that includes the legislative debates to determine the purpose or problem the legislature sought to address.

Administrative Rules and Orders

With the tremendous growth of administrative agencies, Congress and state legislatures have delegated semi-legislative and judicial tasks to a multitude of boards, commissions, and agencies. The legislative body, after identifying a problem, then delegates to the administrative body (within broad guidelines) the authority to issue rules (regulations) and orders (decisions) concerning approved and proscribed conduct. Adoption of zoning regulations by a zoning board is an example. The subject of administrative agencies is discussed in chapter 14.

Court Decisions

Courts contribute to the body of law in three ways: (1) They determine the meaning of constitutions, statutes, and administrative rules and orders; (2) they determine the law where no other source of law is applicable; and (3) they review the constitutionality of and compliance with legal doctrine.

In the United States, all decisions of the U.S. Supreme Court, state supreme courts, and U.S. appellate courts are reported as written official opinions. Likewise, most of the opinions of the U.S. district courts and various state lower courts also are reduced to written reported decisions. In addition, a vast number of administrative agencies prepare written decisions of their cases and proceedings, which are regularly published. To discover how any law applies to a particular problem, it is necessary to find and analyze cases that arose involving similar circumstances. "Case law" is the aggregate of the points of law derived from reported decisions and is separate and distinct from statutory law promulgated by Congress or a state legislature. Dicta (technically called *obiter dicta*) is a term used to describe extraneous comments by the court in its opinion that are not necessary for the decision. Similar to evidence, a prior judicial opinion is only binding to the extent that it is relevant to the case being considered by the court. So, when the court's opinion touches on points not at issue, those extraneous statements may be indications of the court's attitude, but these comments are not considered a binding precedent.

Because there is no unified codification of Common Law, it is necessary to look to the court records of actual decisions to determine the application of Common Law principles to a given question with respect to a specific set of facts. This requires analysis of cases with similar fact situations or presenting similar issues. Since each case or fact situation is unique, the decisions reached in prior cases are valuable as precedents or guidelines. A judicial opinion represents a synthesis of many factors such as a careful examination of the facts, the weight of relevant precedent, the strength of the legal principles involved, and the policy considerations resulting from balancing legitimate competing interests.

Court and administrative decisions are now collected and published by several commercial publishers. Through the use of digests and now by means of computer research, it is possible to find all the relevant cases dealing with similar points of law. One of the most popular systems is prepared by West Publishing Company, which maintains a standard method of reporting such cases. For example, *Brown v. Board of Education*, 347 U.S. 497, 74 S.Ct. 686 (1954), is a U.S. Supreme Court case involving the named parties decided in 1954; the case is found in volume 347 on page 497 of the *United States Reports* system, an official publication of the U.S. government, or in volume 74 on page 686 of West's *Supreme Court Reporter* system. Another case with a citation of 156 P.2d 553, would be a case arising in a state of the Pacific region (e.g., California, Oregon, Washington, Nevada, and Arizona) and found in volume 156 of the second series of West's *Pacific Reporter* system on page 553 as well as in the respective state reports. These legal reports are found in law and many public libraries.

2.5 JUDICIAL PROCESS

At the outset, not only may a brief overview of the method in which the litigation proceeds in the judicial system be useful but also it may be helpful to understand some of the factors that influenced the court's decision and what analysis and reasoning the court used.

The person who institutes a legal proceeding is usually referred to as "plaintiff" or "complainant." The person against whom the judicial proceeding is brought is usually referred to as the "defendant" or the "respondent." The terminology of some appellate cases is confusing because at times the court refers to the appealing party as "appellant" or "appellant in error" and the opposing party as "respondent" or sometimes as "appellee in error." These latter terms arise as the result of giving a new identification to the parties based on which party appeals or seeks relief in the appellate system from the trial or lower court decision.

Legal proceedings are generally commenced by the plaintiff preparing and filing with an officer or clerk of the court an instrument variously referred to as a complaint, charge, or petition. This document briefly describes the breach of a duty by the defendant or the conduct that gives rise to the cause of action entitling plaintiff to a remedy. This instrument is served on the defendant by means of a summons issued by the court or some other means for the purpose of giving notice of the claim to the defendant. Then the defendant is required to respond by filing a document, usually called an answer, in which the defendant sets forth the defense to the claim (e.g., denying the truth of certain allegations or stating a different view of the applicable law). Failure of the defendant to respond within a prescribed time limit results in a default judgment for the plaintiff. After the complaint and answer are filed, the issues formed by these instruments are the basis for trial. At the trial, each of the parties will be given an opportunity to introduce oral and documentary evidence, to cross examine witnesses, and to offer rebuttal evidence to prove their contentions in the case. The trier of fact (court, jury, or administrative agency) will have to resolve conflicts of testimony to determine the true facts.

In other cases, the issue may not turn on specifically conflicting facts but upon an interpretation of the law. There are a clear criteria of priorities where the case is decided on conflicting sources of law. The constitution prevails over statutes. So where the statute violates a constitutional right or guarantee, the statute will fall as null and void. In the next tier, a statute prevails over case law. In turn, dicta of a case (language in the decision that is not essential to the decision) is of lesser weight than the rule of law in the case and dicta is more significant than mere argument.

Supreme Court Justice Benjamin Cardozo, in *The Nature of the Judicial Process*, published in 1921 by Yale University Press, New Haven, Conn., suggested four forces influencing judges in the rationale of their decisional process. These are identified as "logic," "history," "custom," and "utility."

The basis of logic is the use of analogy to reason from one case to another similar case, which is at the heart of the Common Law *stare decisis* doctrine. This use of logic provides the desired goal of stability and predictability. Since history and custom have played fundamental roles in business conduct, law looks to the same business practice in determining expectations and norms of conduct. The force of utility represents factors of justice, ethics, and social welfare. The extent to which the court seeks to be an instrument of social, political, or economic movement depends on the judge's view of the court's role in the judicial system. The activist philosophy is embraced by those judges who believe the judiciary should play a role in policy formulation.

2.6 UNITED STATES LEGAL SYSTEM

The United States has a pluralistic court system consisting of 51 sovereign systems, the federal system and 50 state systems. Although a great diversity exists in the court systems among the states and between the state and federal system, there are two basic types of courts, trial courts and appellate courts. Both the federal and state systems are pyramidal in structure. At the apex of the federal pyramid stands the Supreme Court of the United States, the highest court in the land. This pyramidal organization serves two purposes. First, a supreme court or an appellate court corrects errors made by the trial court.

Second, the higher courts assure uniformity of decision by reviewing cases in which the opinions of two or more lower courts reached different decisions. Trial courts hear and decide controversies by determining facts and applying the appropriate rules of law. Appellate courts review the decisions of trial courts and those of lower appellate courts.

Federal Court System

The federal courts of the United States are created by Congress under the authority of the U.S. Constitution. Their authority is limited by the powers given the federal government in the U.S. Constitution. Article III, Section 1 of the Constitution provides "The judicial power of the United States shall be vested in one Supreme Court and such inferior courts as Congress may from time to time ordain and establish." The Supreme Court (1) has original jurisdiction in cases involving ambassadors, consuls, or where a state is a party and (2) has appellate jurisdiction in cases from federal courts of appeal and state supreme courts where a federal question is involved (rights granted by the U.S. Constitution, laws, or treaties). U.S. courts of appeal were established in 1891 because of the heavy burden of appeals to the Supreme Court. Eleven of these courts (sometimes referred to as circuit courts) exist throughout the United States, and each circuit, except in the District of Columbia, consists of a geographical area of several states. For example, the tenth circuit consists of the midwestern states of Oklahoma, Kansas, Colorado, New Mexico, Utah, and Wyoming. These courts function only as appellate courts, do not conduct trials, and normally are responsible for appeals from federal administrative agencies and conflicts in the law of the lower federal district courts.

The trial courts in the federal system are known as district courts. There are approximately 100 districts throughout the United States so comprised that no state is without at least one district court. The volume of litigation determines the location of federal district courts and the number of judges assigned to any district.

State Court System

In every state there is one court of final appeal, which is commonly called the supreme court such as the Texas Supreme Court. In addition, some states have a system of intermediate appellate courts. The state trial courts have a generally broad power to decide almost every type of case, subject only to the limitations of the constitution and state law. State and local courts are located in nearly every town and county and are the tribunals with which citizens have most frequent contact.

In the state system there is usually a trial court of general jurisdiction over civil, criminal, and equity matters. The nomenclature in the state systems is not uniform, since the trial court is variously called district, circuit, or superior court. Some states also have county courts with geographic jurisdiction restricted to the individual county and with subject matter jurisdiction restricted to wills, trusts, estates, and domestic matters. In addition, some states have small claims courts, justices of the peace, and municipal courts with monetary limits of a few hundred dollars.

2.7 JURISDICTION AND VENUE

The term "jurisdiction" has a technical legal meaning referring to the power of the court under the constitution and enabling legislation (1) to adjudicate concerning the subject matter and the parties, (2) to resolve the dispute, and (3) to render a judgment and to carry the judgment into effect. In some cases, jurisdiction over the subject matter and the parties may be exclusive or concurrent. For example, "county courts" or "probate courts" usually have exclusive jurisdiction over estates of deceased persons. On the other hand, various minor civil damage suits may be filed in either a state small claims court (if the amount does not exceed the court's monetary limits) or in the so-called state court of general jurisdiction, the district court.

Jurisdiction has two aspects, jurisdiction over the subject matter and jurisdiction over the person. Jurisdiction over the subject matter means that the dispute involves a matter that the court has power to hear. For example, the "probate court" does not have jurisdiction over crimes or breaches of contract. In federal courts, jurisdiction over subject matter in a civil action is primarily based on what is referred to as "diversity" or "a federal question." For example, in a civil action, the United States district courts have jurisdiction only when the matter in controversy exceeds the sum or value of $50,000, exclusive of costs and interest, and is based on diversity of citizenship or involves a question of federal law. Diversity of citizenship exists in suits between citizens of different states, a citizen of a state and a citizen of a foreign country, and a state and citizens of another state. For diversity of citizenship to exist, the plaintiff or all of the plaintiffs, if there is more than one, must be citizens of a different state than any of the defendants. For example, a nonresident motorist (resident of Kansas and driving in New Jersey) who injured a pedestrian (resident of New Jersey) may be sued by the injured pedestrian in the state court (New Jersey) where the injury occurred, but the nonresident defendant may remove the case to the federal district court in that state if the monetary amount exceeds the $10,000 requirement. Diversity jurisdiction is based on the theory

of eliminating any prejudice that may exist against "foreign" nonresi-
dents in one party's state court. For the purposes of a law suit in
federal court, a corporation is considered a citizen of the state in which
it is incorporated and of the state in which it has its principal place of
business. As a result, in a law suit against a corporation, federal diversity jurisdiction fails if any one of the parties on the plaintiff side is a
citizen either of the state in which the corporation is chartered or has
its principal place of business.

   Jurisdiction over the person refers to the power to bring the parties to the litigation (i.e., the plaintiffs and defendants) within the
authority of the court. Jurisdiction over the defendant is obtained by
the service of a summons that issues from the court (or publication in
appropriate cases) for the case to be tried. Originally, a court was
limited to a geographical area such as a state, but that has been extended because of modern communication and transportation. Special
rules apply in connection with the use of the highway, and frequently
the statute provides that a nonresident motorist will be deemed to have
appointed an official of the state, usually the secretary of state, as an
agent for service of process. More recent statutes, as of 1988, have extended the jurisdiction of the courts to cover injuries caused by a nonresident "doing business" in the state. This may subject a nonresident
individual or corporation to a law suit if there have been minimal contacts within the state requiring the nonresident to respond if the law
suit does not offend traditional notions of fair play. In addition to jurisdiction over the subject matter and the person, there is the question of
"venue," which refers to the location within the state where the trial
will take place. While these matters are determined by state statute,
customarily suits involving land have venue at the place the land is
situated; in most other suits, the venue is at the location where the
contract was to be performed, where the tort (negligent injury) occurred, or where one of the defendants resides. Venue may be changed
to accommodate the defendants (a more convenient forum) or in order
to obtain a fair and impartial trial.

SIERRA CLUB v. MORTON, SECRETARY OF THE INTERIOR, et al.
405 U.S. 727, 92 S. Ct. 1361 (1972)

   Mr. Justice Stewart: The Mineral King Valley is an area of great
natural beauty nestled in the Sierra Nevada Mountains in Tulare
County, California, adjacent to Sequoia National Park. It has been part

of the Sequoia National Forest since 1926, and is designated as the national game refuge by special Act of Congress. Though once the site of extensive mining activity, Mineral King is now used almost exclusively for recreational purposes. Its relative inaccessibility and lack of development have limited the number of visitors each year, and at the same time have preserved the valley's quality as a quasi-wilderness area largely uncluttered by the products of civilization.

The United States Forest Service, which is entrusted with the maintenance and administration of national forests, began in the late 1940s to give consideration to Mineral King as a potential site for recreational development. Prodded by a rapidly increasing demand for skiing facilities, the Forest Service published a prospectus in 1965, inviting bids from private developers for the construction and operation of a ski resort that would also serve as a summer recreation area. The proposal of Walt Disney Enterprises, Inc., was chosen from those of six bidders, and Disney received a three-year permit to conduct surveys and explorations in the valley in connection with its preparation of a complete master plan for the resort.

The final Disney plan, approved by the Forest Service, outlines a $35 million complex of motels, restaurants, swimming pools, parking lots and other structures designed to accommodate 14,000 visitors daily. This complex is to be constructed on 80 acres of the valley floor under a 30-year use permit from the Forest Service. Other facilities, including ski lifts, ski trails, a cog-assisted railway, and utility installations are to be constructed on the mountain slopes and in other parts of the valley under a revocable special-use permit. To provide access to the resort, the State of California proposes to construct a highway 20 miles in length. A section of this road would traverse Sequoia National park, as would a proposed high-voltage power line needed to provide electricity for the resort. Both the highway and the power line require the approval of the Department of the Interior, which is entrusted with the preservation and maintenance of the national parks.

Representatives of the Sierra Club, who favor maintaining Mineral King largely in its present state, followed the progress of recreational planning for the valley with close attention and increasing dismay. They unsuccessfully sought a public hearing on the proposed development in 1965, and in subsequent correspondence with officials of the Forest Service and the Department of the Interior, they expressed the Club's objections to Disney's plan as a whole and the particular features of it. In June 1969, the Club filed the present suit in the United States District Court for the Northern District of California, seeking a declaratory judgment that various aspects of the proposed development contravene federal regulations governing the preservation of national parks, forests, and game refuges, and also seeking preliminary and per-

manent injunctions restraining the federal officials involved from granting their approval or issuing permits in connection with the Mineral
King project. The petitioner Sierra Club sued as a membership corporation with "a special interest in the conservation and the sound maintenance of the national parks, game refuges and forests of the country,"
and invoked the judicial-review provisions of the Administrative Procedure Act.

   After two days of hearings, the District Court granted the requested preliminary injunction. It rejected the respondents' challenge to
the Sierra Club's standing to sue; and determined that the hearing has
raised questions "concerning possible excess of statutory authority, sufficiently substantial and serious to justify a preliminary injunction. . . ."
Respondents appealed, and the Court of Appeals reversed 433 F.2d 24
(9th Cir. 1970). With respect to petitioner's standing, the court noted
that there was "no allegation in the complaint that members of the
Sierra Club would be affected by the actions of (the respondents) other
than the fact that the actions are personally displeasing or distasteful
to them."

   The first question presented is whether the Sierra Club has alleged facts that entitle it to obtain judicial review of the challenged action. Whether a party has a sufficient stake in an otherwise justiciable
controversy to obtain judicial resolution of that controversy is what has
traditionally been referred to as the question of standing to sue. Where
the party does not rely on any specific statute authorizing invocation of
the judicial process, the question of standing depends upon whether the
party has alleged such a "personal stake in the outcome of the controversy," as to ensure that "the dispute sought to be adjudicated will
be presented in an adversary context and in a form historically viewed
as capable of judicial resolution." Where, however, Congress has
authorized public officials to perform certain functions according to law,
and has provided by statute for judicial review of those actions under-
certain circumstances, the injury as to standing must begin with a
determination of whether the statute in question authorizes review at
the behest of the plaintiff.

   The Sierra Club relies upon Section 10 of the Administrative Procedure Act (APA). . . which provides:

> A person suffering legal wrong because of agency action; or adversely af-
> fected or aggrieved by agency action within the meaning of a relevant
> statute, is entitled to judicial review thereof.

   Early decisions under this statute interpreted the language as
adopting the various formulations of "legal interest" and "legal wrong"
then prevailing as constitutional requirements of standing.

The injury alleged by the Sierra Club will be incurred entirely by reason of the change in the uses to which Mineral King will be put, and the attendant change in the aesthetics and ecology of the area. Thus, in referring to the road to be built through Sequoia National Park, the complaint alleged that the development "would destroy or otherwise adversely affect the scenery, natural and historic objects and wildlife in the park and would impair the enjoyment of the park for future generations." We do not question that this type of harm may amount to an "injury in fact" sufficient to lay the basis for standing under Section 10 of the APA. Aesthetic and environmental well-being, like economic well-being, are important ingredients of the quality of life in our society, and the fact that particular environmental interests are shared by the many rather than the few does not make them less deserving of legal protection through the judicial process. But the "injury in fact" test requires more than an injury to a cognizable interest. It requires that the party seeking review be himself among the injured.

The impact of the proposed changes in the environment of Mineral King will not fall indiscriminately upon every citizen.The alleged injury will be felt directly only by those who use Mineral King and Sequoia National Park, and for whom the aesthetic and recreational values of the area will be lessened by the highway and ski resort. The Sierra Club failed to allege that it or its members would be affected in any of their activities or pastimes by the Disney development. Nowhere in the pleadings or affidavits did the Club state that its members use Mineral King for any purpose, much less that they use it in any way that would be significantly affected by the proposed actions of the respondents.

The Club apparently regarded any allegations of individualized injury as superfluous, on the theory that this was a "public" action involving questions as to the use of natural resources, and that the Club's long-standing concern with and expertise in such matters were sufficient to give it standing as a "representative of the public." This theory reflects a misunderstanding of our cases involving so-called "public actions" in the area of administrative law.

The trend of cases arising under the APA and other statutes authorizing judicial review of federal agency action has been toward recognizing that injuries other than economic harm are sufficient to bring a person within the meaning of the statutory language, and toward discarding the notion that an injury that is widely shared is ipso facto not an injury sufficient to provide the basis for judicial review. We noted this development with approval in *Data Processing*, 397 U.S. 150, 154, in saying that the interest alleged to have been injured may reflect "aesthetic, conservational, and recreational" as well as "economic values." But broadening the categories of injury that may be alleged in support of standing is a different matter from abandoning the requirement that the party seeking review must himself have suffered an injury.

It is clear that an organization whose members are injured may represent those members in a proceeding for judicial review. But a mere "interest in a problem," no matter how long-standing the interest and no matter how qualified the organization is in evaluating the problem, is not sufficient by itself to render the organization "adversely affected" or "aggrieved" within the meaning of the APA. The Sierra Club is a large and long-established organization, with a commitment to the cause of protecting our Nation's natural heritage from man's depredations. But if a "special interest" in this subject were enough to entitle the Sierra Club to commence this litigation, there would appear to be no objective basis upon which to disallow a suit by any other bona fide "special interest" organization, however small or short-lived. And if any group with a bona fide "special interest" could initiate such litigation, it is difficult to perceive why any individual citizen with the same bona fide special interest would not also be entitled to do so.

The requirement that a party seeking review must allege facts showing that he is himself adversely affected does not insulate executive action from judicial review, nor does it prevent any public interests from being protected through the judicial process. It does serve as at least a rough attempt to put the decision as to whether review will be sought in the hands of those who have a direct stake in the outcome. That goal would be undermined were we to construe the APA to authorize judicial review at the behest of organizations or individuals who seek to do no more than vindicate their own value preferences through the judicial process.

The principle that the Sierra Club would have us establish in this case would do just that.

[W]e conclude that the Court of Appeals was correct in holding that the Sierra Club lacked standing to maintain this action . . . . The judgment is affirmed.

Mr. Justice Douglas, dissenting. I share the view of my Brother Blackmun and would reverse the judgment below.

The critical question of "standing" would be simplified and also put neatly in focus if we fashioned a federal rule that allowed environmental issues to be litigated before federal agencies or federal courts in the name of the inanimate object about to be despoiled, defaced, or invaded by roads and bulldozers and where injury is to the subject of public outrage. Contemporary public concern for protecting nature's ecological equilibrium should lead to the conferral of standing upon environmental objects to sue for their own preservation. See Stone, *Should Trees Have Standing?—Toward Legal Rights for Natural Ob-*

jects, 45 S. Cal. L. Rev. 450 (1972). This suit would therefore be more properly labeled as *Mineral King v. Morton.*

Inanimate objects are sometimes parties in litigation. A ship has a legal personality, a fiction found useful for maritime purposes. The corporation sole, a creature of ecclesiastical law, is an acceptable adversary and large fortunes ride on its cases. The ordinary corporation is a "person for purposes of the adjudication processes, whether it represents proprietary spiritual, aesthetic or charitable causes."

So it should be as respects valleys, alpine meadows, rivers, lakes, estuaries, beaches, rivers, groves of trees, swampland, or even air that feels the destructive pressures of modern technology and modern life. The river, for example, is the living symbol of all the life it sustains or nourishes—fish, aquatic insects, water fowls, otter, deer, elk, bear, and all other animals, including man, who are dependent on it or who enjoy it for its sight, its sound, or its life. The river as plaintiff speaks for the ecological unit of life that is part of it. Those people who have a meaningful relation to that body of water—whether it be fisherman, a canoeist, a zoologist, or a logger—must be able to speak for the values which the river represents and which are threatened with destruction.

I do not know Mineral King. I have never seen it nor traveled it, though I have seen articles describing its proposed "development" notably Hano, "Protectionists vs. Recreationists—The Battle of Mineral King," *N.Y. Times Mag.,* Aug. 17, 1969, p. 25. The Sierra Club in its complaint alleges that "[o]ne of the principal purposes of the Sierra Club is to protect and conserve the national resources of the Sierra Nevada Mountains." The District Court held that this uncontested allegation made the Sierra Club "sufficiently aggrieved" to have "standing" to sue on behalf of Mineral King.

Mineral King is doubtless like other wonders of the Sierra Nevada such as Tuoluone Meadows and the John Muir trail. Those who hike it, fish it, hunt it, camp in it, frequent it, or visit it merely to sit in solitude and wonderment are legitimate spokesmen for it, whether they may be few or many. Those who have that intimate relation with the inanimate object about to be injured, polluted, or otherwise despoiled are its legitimate spokesmen.

The Solicitor, whose views on this subject are in the Appendix to this opinion, takes a wholly different approach. He considers the problem in terms of "government by the Judiciary." With all respect, the problem is to make certain that the inanimate objects, which are the very core of America's beauty, have spokesmen before they are destroyed.

Yet the pressures on agencies for favorable action one way or the other are enormous. The suggestion that Congress can stop action which is undesirable is true in theory; yet even Congress is too remote to give meaningful direction and its machinery is too ponderous to use very often. The federal agencies of which I speak are not venal or corrupt. But they are notoriously under the control of powerful interests who manipulate them through advisory committees, or friendly working relations, or who have that natural affinity with the agency which in time develops between the regulator and the regulated.

The Forest Service—one of the federal agencies behind the scheme to despoil Mineral King—has been notorious for its alignment with lumber companies, although its mandate from Congress directs it to consider the various aspects of multiple use in its supervision of the national forests.

The voice of the inanimate object, therefore, should not be stilled. That does not mean that the judiciary takes over the managerial functions from the federal agency. It merely means that before these priceless bits of Americana (such as a valley, an alpine meadow, a river, or a lake) are forever lost or are so transformed as to be reduced to the eventual rubble of our urban environment, the voice of the existing beneficiaries of these environmental wonders should be heard.

Perhaps they will not win. Perhaps the bulldozers of progress will plow under all the aesthetic wonders of this beautiful land. That is not the present question. The sole question is, who has the standing to be heard?

UNITED STATES v. BISHOP PROCESSING COMPANY
287 F.Supp. 624 (D.C. Md. 1968),
Cert. den. 398 U. S. 904 (1970)

Thomsen, Chief Judge. In this action, the government seeks under the Clean Air Act to enjoin Bishop Processing Company [defendant], the operator of a rendering and animal reduction plant near Bishop, Worcester County, Maryland, from discharging malodorous air pollutants, which it is alleged, move across the state line and pollute the air in and around Selbyville, Delaware.

The first ground stated in defendant's motion to dismiss is that the Clean Air Act is an unconstitutional attempt by Congress to control purely local intrastate activities over which Congress has no power to legislate. Defendant argues (a) that the movement of pollutants across state lines is not interstate commerce itself, and (b) has no substantial effect on interstate commerce.

(a) The movement of pollutants across a state line is a proper jurisdictional basis for the provisions of the Act relating to the abatement of interstate air pollution. Such movement of pollutants across state lines constitutes interstate commerce subject to the power granted to Congress by the Constitution to regulate such commerce. Whether the originator of the pollution directs it across the state border intentionally is immaterial. In *Thornton v. United States* (1926), the owner of cattle which ranged on land near the Florida-Georgia border claimed that they were not within interstate commerce and, consequently, that he could not be required to comply with a federal requirement for the inspection and preventive treatment of cattle in an area under quarantine. The Supreme Court upheld the constitutionality of the applicable statute, stating:

> [I]t is said that these cattle do not appear to have been intended to be transported by rail or boat from one state to another and this only is interstate commerce in cattle under the Constitution. They were on the line between the two states. To drive them across the line would be interstate commerce, and the act of 1905 expressly prohibits driving them on foot when carrying contagion. It is argued, however, that when the cattle only range across the line between the states and are not transported or driven, their passage is not interstate commerce. We do not think that such passage by ranging can be differentiated from interstate commerce. It is intercourse between states, made possible by the failure of owners to restrict their ranging and is due, therefore, to the will of their owners.

In *United States v. Darby*, 61 S. Ct. 451 (1941), the Court quoted as follows:

> The power of Congress over interstate commerce is complete in itself, may be exercised to its utmost extent, and acknowledges no limitations other than are prescribed in the Constitution. It is no objection to the assertion of the power to regulate interstate commerce that its exercise is attended by the same incidents which attend the exercise of the police power of the states.

The commerce power may be exercised to achieve socially desirable objectives, even in the absence of economic considerations. *Brooks v. United States* 45 S. Ct. 345 (1959).

(b) Defendant contends that pollution has no substantial and harmful effects on commerce, arguing that the congressional finding that air pollution has resulted in hazards to air and ground transportation is clearly erroneous, and that if pollution has any effect on air and ground transportation, such effect has been isolated and insubstantial.

Since the provisions of the Act relating to the abatement of interstate air pollution may properly be based on the interstate movement of the pollutants themselves, it is not necessary that such pollutants interfere with interstate commerce in order to sustain this exercise of the commerce power. Congress, however, concluded that:

> the growth in the amount and complexity of air pollution brought about by urbanization, industrial development, and the increasing use of motor vehicles, has resulted in mounting dangers to the public health and welfare, including injury to agricultural crops and livestock, damage to and the deterioration of property, and hazards to air and ground transportation.

The finding in [this section of the statute] is adequately supported by the legislative history.

A court's review of such a congressional finding is limited. The only questions are whether Congress had a rational basis for finding that air pollution affects commerce, and if it had such a basis, whether the means selected to eliminate the evil are reasonable and appropriate.

Defendant argues that the congressional finding that air pollution has an effect on air and ground transportation is "clearly erroneous," since the legislative history provides only "isolated and insubstantial" interferences with transportation. The power of Congress to regulate activities affecting interstate commerce is to be determined not only by quantitative effect of individual operations, but also by the total effect of many individual interferences with commerce.

Defendant next argues that non-visible pollution has no effect upon commerce and consequently Congress has no authority over "odorous pollution."

The complaint recites the following findings of fact made by the Hearing Board:

> 3. The malodorous pollution consists of sickening, nauseating and highly offensive odors which are pervasive in effect to the interstate Selbyville, Delaware area. Such noxious, malodorous air pollution endangers the health and welfare of persons in the town of Selbyville, Delaware and adjacent and contiguous areas. It causes nausea, sleeplessness, and revulsion, thereby imposing a physiological and psychological burden on persons subjected thereto; and it adversely affects business conditions and property values and impedes industrial development.

Malodorous pollution which "adversely affects business conditions and property values and impedes industrial development" would clearly interfere with interstate commerce.

Defendant's argument that there is no economic relationship be-tween the activity regulated and the commerce protected must also fail. Congress undertook to regulate movement of pollutants across state borders, and it is alleged that those pollutants do interfere with inter-state commerce. Hence, the "local activity" (the operation of the render-ing plant) is subject to the power of Congress to regulate interstate commerce. In *Heart of Atlanta Motel, Inc. v. United States*, 85 S. Ct. 348, 358 (1964) the Court said:

> It is said that the operation of the motel here is of a purely local charac-ter. But, assuming this to be true, if it is interstate commerce that feels the pinch, it does not matter how local the operation which applies the squeeze.

The court added:

> Thus the power of Congress to promote interstate commerce also includes the power to regulate the local incidents thereof, including local activities in both the States of origin and destination, which might have a substan-tial and harmful effect upon that commerce.

Congress had a rational basis for finding that air pollution affects commerce, and the means selected by Congress to eliminate the harm-ful effects of the interstate movement of air pollutants are reasonable and appropriate.

DISCUSSION QUESTIONS

1. What is the rationale of the majority opinion in the *Morton case* as to legal necessity of standing to sue? How does the dissenting opinion deal with these arguments? If "standing" is a threshold question for the court to decide at the outset of the trial, does the court's assessment of the merits of the plaintiff's case enter into that determination?

2. In the *Bishop Processing case,* jurisdiction for federal authority to enforce the Clean Air Act is based on the commerce clause of the U.S. Constitution. What are the limits of the federal government's authority to regulate inter-state commerce? Does the court's definition of commerce square with the generally accepted usage of that term? What are the limits of the court in considering the means which the legislature used in regulating a particular activity?

3. Define and explain (1) statutory law, (2) Common Law, and (3) equity. Where do the courts get authority to interpret the Common Law?

4. Would you consider a ruling excluding certain evidence as "substantive" or "procedural" law?

5. How are conflicts between a law and the constitution resolved?

6. Why are there distinct and separate federal and state court systems? What is the jurisdiction of the federal court system, and what kinds of cases does it hear? What is the jurisdiction of the state court system and what kinds of cases does it hear?

3

DEFINITION, FORMALITY, AND NATURE OF CONTRACT

3.1 INTRODUCTION

The civilized world depends on the legal recognition of contractual obligations. The faithful performance of agreements is an essential element of not only the business world but also for planning by individuals. Contracts involve the purchase of materials and services for business as well as the purchase of personal items, such as gasoline or cable television. Turning on and supplying electricity at a construction site is performed under the terms of a contract. National defense also requires a series of contracts to procure hardware and services, and even foreign treaties represent a form of contractual commitment at the sovereign government level.

Contracts may lap over into other areas of the law. For example, when a car is left at the service station for repair, there is a contract for the work to be done, and also a contract such as a bailment (discussed in chapter 11) as to the care of the car while it is in the service station's possession. In addition, a contractual matter develops when the mechanic or service writer discusses the work to be performed since that person may be a representative (such as an agent or partner) of the owner of the garage.

All contracts are agreements, and, while all agreements, ethically, at least, may be considered obligations, these ethical obligations may not always be enforceable at law. In this chapter the focus is primarily on legal contracts, but those that fail to meet that criteria may nevertheless be ethical obligations only, without direct legal enforcement. For example, if X, an employer, agrees to inform Y, an applicant for an engineering job, when X later employs additional engineers and indeed fails to notify Y, X may have breached an ethical duty but not a legal contract. Further study will illustrate the reasons for this distinction. Accordingly, the agreement to notify Y when X is hiring is not an enforceable contract for which Y could recover monetary damages caused by X's failure to inform Y. On the other hand, if X and Y entered into an agreement that X would rent Y's car on Thanksgiving Day for $150, and X failed to rent the car, that agreement would be a contract and capable of enforcement in the courts. Each party to the contract would have a legal action available if the other failed to perform as agreed.

Many of the rules of contract law have their roots in principles developed in the 18th and 19th centuries. The social conditions existing at that time played an important role in shaping the development of contract law. At that time, most dealings were on a face-to-face basis and with the parties acquainted with each other or at least knowing their reputation for fair dealing. However, with the advent of the so-called Industrial Revolution, business transactions became more complex and less personalized. As the feudal system of self sufficiency disappeared and an economic system in which concepts of division of labor and exchanges of goods and services developed, the early judicial concept of "contract" was transformed to a practical tool of merchants. Economic theories of *laissez-faire* (free market) served as an important public policy in shaping commercial transactions and law. Thus a practice developed of placing in contracts express disclaimers for any liability for injury caused by the merchant's products, and these provisions were enforced against the purchaser.

With mass production and enlarged impersonal distribution systems, the legislature responded to the growth of consumerism. Likewise, the legal system reacted to this industrial development by modifications in contract law in the process. The result was a shift in emphasis from protecting business and promoting industrialization to protecting consumers and workers.

An underlying concept of contract law remains in the idea that a contract is an agreement freely entered into by the parties. However, the policy trend is for more public input into contracts with the courts viewing with great suspicion clauses that fail to consider the bargaining power of the parties or attempts by manufacturers to limit, by contractual clauses, responsibility for their products.

The Uniform Commercial Code (UCC) prepared by the American Law Institute and the National Conference of Commissioners on Uniform Laws has been adopted by all states except Louisiana. The purpose was to establish a uniform system of law for commercial transactions that often are conducted across state lines. Article 2 of the UCC applies to "transactions in *goods*" (emphasis added). Frequently, the issue arises as to whether the contract is governed by Article 2 when it involves goods and services, since contracts solely for services, are not within the scope of Article 2. There is authority that if an integral part of the contract concerns the furnishing of goods, and the cause of action allegedly arises from the goods portion of the contract, Article 2 should apply even though the services portion of the contract dominates the transaction.

3.2 DEFINITION, FORMALITY, AND NATURE OF CONTRACT

An agreement that the law recognizes as a contract is defined in *Black's Law Dictionary* as "a promissory agreement between two or more parties which creates, modifies, or destroys a legal obligation." Sometimes, a contract is simply defined as "an agreement enforceable at law." This distinction requires that the agreement rising to the level of an enforceable contract must have definite elements and meet certain conditions.

The RESTATEMENT of Contracts defines a contract "as a promise or set of promises for the breach of which the law gives a remedy or the performance of which the law recognizes as a duty." This definition is circular, since defining a contract as a legally enforceable promise begs the issue by stating that a promise is only legally enforceable if it is a contract. Under Section 1-201(11) of the Uniform Commercial Code, "contract" means the total legal obligation that results from the parties' agreement as affected by the Code and applicable case law. The term "agreement" means the bargain of the parties arising from their language or by implication of other circumstances, including their course of dealing or the usage of their trade. A contract is not a "law," but it is the agreement plus the legal doctrine that makes a contract an enforceable obligation.

Promise

The word "promise" in ordinary usage has various shades of meaning. However, in both popular and legal parlance, a promise is assurance that certain action or conduct will or will not occur. While the underlying concept of every contract is a promise, express or implied, no particular words of art are necessary to create a promise. The essential element is a fair interpretation of the words or action that make it appear that a promise was intended. A mere intention or desire must be distinguished since a promise constitutes an undertaking to carry the intention into effect. In *Webster v. Upton*, 91 U.S. 65, 23 L.Ed. 384 (1875), the court stated "No form of words is necessary to constitute a promise binding in law; an implied promise may be shown by proof of circumstances that show that the party intended to assume an obligation by putting himself into a position which requires the performance of duties."

Consideration

The binding character of an agreement is the essence of a contract. In the Roman legal system, formalities such as requiring the instrument to be in writing or under a "seal" supplied the ingredient to assure the parties' acceptance of that contractual quality. With the recognition that many transactions did not lend themselves to such formality, the Common Law devised a more workable element referred to as "consideration," which nevertheless was adequate to demonstrate the binding intention of the parties. Consideration is defined as a legal detriment suffered by the promisee that is requested by the promisor in exchange for promisor's commitment. Usually in bilateral contracts, this takes the form of mutual promises; however, in addition to a promise, consideration may consist of an act or a forbearance (giving up a valid claim such as releasing a right to sue). Ordinarily, the court does not examine the adequacy of the consideration but generally only determines whether the person against whom the promise is to be enforced (promisor) received something in return from the person to whom the promise was made (promisee).

Express and Implied Contracts; Quasi-Contract

Contracts are frequently classified as express or implied contracts. An express contract is one in which the terms of the agreement are fully and openly incorporated into the terms of the agreement at the time the parties entered the contract. *McDonald v. Thompson*, 22 S. Ct. 297 (1902). An implied contract is inferred from the facts and circumstances of the case and not found from formal or explicit words. An implied contract is raised only when the facts are such that an intent may be fairly inferred by the conduct of both parties and will not be implied against either the express declaration of the party to be charged or against the understanding of the parties. In other words, an implied contract will not arise where the facts are wholly inconsistent with the contract to be implied.

In *Woolley v. Hoffman-LaRoche, Inc.*, 491 A.2d 1257 (1985), involving dismissal of an engineer, employee-at-will without an employment contract, the court found the employer's action to be a breach of contract, holding that the employment manual's statement that "an employee would be discharged only for cause" was an implied [contract] in the absence of a clear and prominent disclaimer.

Quasi-contracts, as distinguished from contracts implied in fact, are contracts implied in law. Quasi-contracts are generally based on the theory of unjust enrichment, where, in all justice and fairness, it would be wrong to allow one person to receive something of value from another without having to pay for it. In *United States v. Young Lumber Co.*, 376 F. Supp. 1290 (D.S.C. 1974), the court found an implied promise on the part of the recipient to pay the reasonable value for services where the employee of a subcontractor brought equipment he had rented to the job site to replace equipment that a third party had supplied but removed. The subcontractor admitted that the employee's equipment was on the job site and that it had been used but contended there was no express agreement under which the subcontractor would pay the employee rent for the equipment. The court held that where one performs for another a useful service for which there is usually a charge, and that such service is rendered with the knowledge and approval of the recipient who either expresses no dissent or avails himself or herself of the service rendered, the law raises an implied promise. In such situations the contract, implied in law, does not depend on a consensual arrangement, and the recovery is based on a theory of unjust enrichment commonly called "quasi-contract."

Executory and Executed Contracts

An executory contract is one that is yet to be performed, while an executed contract is one that has been fully performed. A contract may be executed as to one party and executory as to the other party. For example, X, an engineer, finished the design and specifications for a particular project and delivered them to Y, the developer who ordered them, although Y has not yet paid the engineer the agreed compensation. As to X, the engineer, the contract is executed and as to Y the contract is executory.

Void, Voidable, and Unenforceable Contracts

A void contract is invalid and produces no legal obligations. In *Bierman v. Hagstrom Construction Company*, 170 P.2d 1138 (Cal. 1959), the court treated the contract as illegal where plaintiff, a painting contractor, was on a job for about 10 months during which time his license was under suspension for 5 weeks because of his failure to pay the $5.00 renewal fee on time. The court denied plaintiff any recovery, reasoning that it was illegal for the plaintiff to perform work as a contractor without a valid contractor's license at all times, and the court refused to lend its support to an illegal contract. This same policy is now incorporated in Section 7031 of the California Business and Professions Code providing that no person acting in the capacity of a contractor may maintain any action in court for the collection of compensation for work done as a contractor without proving that he or she was duly licensed at all times during the performance of the work.

A voidable contract is a contract that is generally unenforceable by one party but enforceable by the other and unless rescinded imposes legal obligations the same as a valid contract. By way of illustration, a minor, who is a party to a contract, may disaffirm the contract prior to reaching majority. The application of such principles to a minor, insane person, or intoxicated person are discussed later in this chapter.

An unenforceable contract may create an obligation, but it cannot be enforced by legal proceedings. Verbal contracts, which are required to be in writing under the Statute of Frauds, and contracts, which are barred by the Statute of Limitations because of the passage of time before a claim is filed, illustrate unenforceable contracts. The Statute of Limitations usually prescribes a designated period of time within which certain claims must be brought. Construction claims present particular problems with respect to the point in time at which the period "begins." The law has had difficulty in selecting from a number of base points, such as the wrongful act, the occurrence of damage or the discovery of the defect. Some jurisdictions look to the completion of the project with respect to design errors, but in *Cubito v. Kreisberg*, 419 N.Y.S. 578 affrm. 415 N.E.2d 979 (1980), it was held that the completion date was a proper basing point in dealing with claims by the client owner. However, the court distinguished contract claims from tort claims brought by injured third parties and held that the basing point for the injured third party was when the wrongful act was discovered, which is the time of injury.

3.3 FORMATION OF CONTRACTS

In the famous Dartmouth College case, *Dartmouth College v. Woodward*, 17 U.S. 518, 4 L.Ed. 629 (1819), the court said

> The ingredients requisite to form a contract are parties, consent, and an obligation to be created or dissolved; these must concur, because the

regular effect of all contracts is on one side to acquire, and on the other to part with some property or rights, or to abridge or restrain natural liberty by binding the parties to do, or restraining them from doing, something which before they might have done, or omitted.

The case law spells out that a valid contract requires that five elements must exist according to law: (1) agreement involving an offer and an acceptance, (2) competent parties, (3) consideration, (4) lawful purpose, and (5) formality. However, in many actual situations, the contract formation is not a concise straightforward offer to be followed by a categorical single word "yes" acceptance. Rather, the murky facts frequently encountered are illustrated in the Western Contracting case, which is presented at the end of this chapter.

The Offer

The touchstone of a contract is the mutual understanding as to what each party will give and receive to perform the contract obligations. Generally, prior to the existence of the contract, the parties will engage in negotiations in an effort to arrive at a mutual understanding. Usually, one party (offeror) will eventually suggest a proposition constituting the offer in a manner manifesting to another party that if the offer is accepted by the other party (offeree) a contract will result.

Intent and Communication

It is essential that the offer be made with the intention to enter into a contract, that it be communicated to the offeree, that it be definite and certain, and that it not be a mere invitation to negotiate. In determining the offeror's intent, the courts will apply the "reasonable man" standard as a means of determining the presence or absence of contractual intent. The offeror's actual state of mind is irrelevant; the critical factor is whether the offeree, as an ordinary, reasonable person, is justified in believing that an offer was made. It follows that statements made in jest or under duress normally will not be considered a valid offer since they lack the contractual intent. In *Barnes v. Treece*, 549 P.2d 1152 (Wash. 1976), the court described the implications of offers made in jest as

> If jest is not apparent and a reasonable hearer would have believed that an offer was being made, the speaker risks the formation of a contract which was not intended.

The offer, whether oral or written, must be communicated to the offeree, and that communication must be as the result of the offeror's action. Accidental or incidental unintended communication to the offeree is not sufficient to constitute an offer.

The actual offer must be distinguished from the preliminary negotiations during which time the parties are bargaining in an effort to arrive at the terms of a definitive agreement. In this regard, advertisements pose special problems as to whether they constitute offers. Usually, advertisements are considered only invitations to negotiate or to make an offer. Where the advertisement, such as either offering to pay a reward or proposing to hold an auction without reservation, contains a positive promise by the advertiser as to what is expected in return, then the advertisement may be construed to be an offer.

In *O'Keefe v. Lee Calan Imports, Inc.*, 262 N.E. 758 (Ill. 1970), involving plaintiff's attempt to purchase a Volvo station wagon based on an erroneous advertisement in the *Chicago Sun Times* that incorrectly advertised a Volvo for $1,095 rather than $1,795, the court held that the advertisement was merely an invitation to negotiate and not an offer unless "special circumstances" were present. The court distinguished the case from an earlier case in which a car dealer's advertisement that purchasers of a 1954 Ford could trade the car for a 1955 Ford at no additional cost was held to be an offer since the advertisement required "the performance of an act" by the purchase of the 1954 Ford.

The bidding process in construction and manufacturing work is a similar troublesome source of legal disputes. Prime contractors or general contractors who advertise for bids usually are held to have made only an invitation for offers and the bids submitting in response are considered as offers. When the job or project is awarded to the general contractor, the general contractor is not required to accept the subcontractor's "bid," although it was used as the basis for the general bid.

Definitiveness of Terms

An offer should be definite and certain as to the essential elements of performance such as time, price, quantity, quality, and identification of subject matter. Certain terms may be supplied by custom, usage, or the law. Therefore, in a contract under which the buyer agrees to purchase buyer's requirements from a seller, and seller agrees to supply all of the needs of the buyer, the UCC Section 2-306 provides

Output, Requirements and Exclusive Dealings. (1) A term which measures the quantity by the output of the seller or the requirements of the buyer means such actual output or requirements as may occur in good faith, except that no quantity unreasonably disproportionate to any stated estimate or in the absence of a stated estimate to any normal or otherwise comparable prior output or requirements may be tendered or demanded.

In *Jansen v. Phillips*, 437 P.2d 189 (Wash. 1968), where plaintiff architect agreed to "substantially landscape most of [defendant's] property for $6,500 or nearly so" and defendant had indicated that he wanted a "first class job," the court indicated that where the parties intended to make a contract the court should tread lightly in concluding that the parties were not successful because of an absence of reasonable certainty. The Common Law rules concerning indefiniteness are severely altered in sales transactions between merchants. UCC Section 2-204(3) provides that a contract for the sales of goods does not fail, even though one or more terms are left open and sets forth a methodology for finding missing terms in sales contracts.

Under the methodology for contracts involving the sale of goods under the UCC, the court is instructed to look to "course of performance," then "course of dealing," and, finally, to "usage or trade." If price and payment terms are missing, UCC Section 2-305 infers "a reasonable price at the time [of] delivery." If the time and place of delivery is missing, UCC Section 2-308 and Section 2-309 indicates "seller's place of business will be imputed as the place of delivery" and "a reasonable time for delivery is implied."

Termination and Duration of the Offer

The offeree has the power to accept the offer at any time before it is terminated. The offer may be terminated by (1) revocation, (2) express terms or implied terms for revocation, (3) operation of law, (4) lapse of time, or (5) destruction of the subject matter. Stated differently, the offer that has been properly communicated to the offeree continues until it lapses or expires, it is revoked by the offeror, it is rejected or accepted by the offeree, or it becomes illegal or impossible by operation of law.

Revocation. An offer may be revoked by the offeror at any time prior to its acceptance notwithstanding stipulations to the contrary. In most states, the revocation must be communicated to the offeree, but the revocation may be communicated to the offeree by any means of notice. For example, the sale of property covered by the offer will constitute notice of revocation (termination) if the offeree has notice of the sale. In a few states, a revocation is effective when placed in the offeror's designated means of communication.

Rejection and counteroffer. While the offeree has no obligation to reply to the offer, the response may be a rejection or a counteroffer. In either case, if received by the offeror, the offer is terminated and cannot be revived by any act on the part of the offeree. When the offeree replies to the offeror by suggesting other terms, a counteroffer is made which may then be accepted by the previous offeror. The counteroffer in effect reverses the offeror/offeree relationship.

Express terms. When the offer contains an express time limitation, the offer will expire automatically by the lapse of the time specified in the offer.

Implied terms. In the absence of express time limitations, an offer expires after a reasonable time has elapsed. "Reasonable time" depends on a variety of factors including the method of communication and the nature of the property or subject matter. By way of illustration, an offer made directly to offeree in a conversation may not extend beyond the conversation, and an offer made by wire may be considered of short duration and require acceptance by wire. Where there is no express stipulation as to the expiration of the offer, an offer to sell securities may lapse sooner than an offer to sell real estate.

Operation of law. Normally, termination of an offer by operation of law involves situations where completion of the making of the contract is rendered impossible without the fault of either the offeror or the offeree. Death or insanity of the offeror will terminate the offer and the offeror, or anyone in behalf of the offeror in case of death or insanity, has no obligation to notify the offeree that the offer is terminated.

Destruction of the subject matter. If the subject matter of the contract is destroyed without the fault of either party, after making the offer but before acceptance, the offer is terminated. So, if X offers to rent office space to Y and the building is destroyed by fire before the offer is accepted, the offer is terminated.

The Acceptance

Since contracts are based on the mutual agreement of the parties, it is a condition prerequisite that the offeree indicate a willingness to be bound by the terms of the offer. Acceptance of an offer is the manifestation by the offeree to comply with the terms of the offer.

Medium of acceptance. To be effective, the acceptance must be communicated to the offeror. Ordinarily, unless the offeror indicates otherwise, the offer may be accepted by any medium. When the offeror has indicated a means of acceptance and unless the offer requires actual receipt by the offeror of the acceptance, the contract is formed when the acceptance is placed in the proper channel of communication such as delivery to the telegraph office or posting in the mail. Under the so-called "Mail Box" rule established by the famous English case of *Adams v. Lindsell*, a mailed acceptance is effective when deposited in the mail in instances where the offeror has requested an acceptance by mail. The acceptance need not be received by the offeror so long as it was properly addressed and dispatched. If the offeror has stipulated that the acceptance must be "received" to be effective, then the contract is not formed until the acceptance is received. Under the mail box rule, the weight of authority holds that an acceptance deposited in the mail is effective even though the offeree, in order to retract the acceptance, intercepts the letter before delivery to the offeror.

Performance as acceptance. As contrasted with bilateral contracts where both parties make promises, unilateral contracts are agreements in which a promise on one side may be accepted by performance of an act on the other side.

Silence as acceptance. The intention to accept the offer must be manifested in some manner and the determination to accept alone is insufficient. In the absence of circumstances that impose a duty to speak or act, silence or inaction will not constitute an acceptance. In circumstances where the parties have had previous dealings or where the offeree accepts and receives the benefits of goods or services, silence or inaction may constitute acceptance.

3.4 CAPACITY OF THE PARTIES

Contractual capacity refers to the ability to perform legally valid acts, to incur legal liability, and to acquire legal rights. The general underpinning of both the offer and the acceptance indicates a contractual relationship between the offeror and the offeree necessitating that both parties have legal capacity to contract.

Minors

Minors are presumed to lack capacity to contract and are allowed to disaffirm contracts. The law makes an exception for "necessities" for which the minor is liable not in contract but in quasi-contract. "Necessities" are those things that are needed for the minor's subsistence, such as food and lodging, medical services, education, and clothing, and the extent of such necessities are measured by age, state, and condition in the life of the minor. Likewise, contracts by insane persons may be disaffirmed if the insane person lacks sufficient mental capacity to understand the nature of the transaction.

If the contract has been fully or partially executed, the minor has the right to disaffirm and obtain return of the consideration. The minor must return the consideration that the minor received based on the equitable principle that those seeking equity must "do equity." This requirement that the minor should place the other party in status quo is becoming more popular under both statutory amendments and judicial thinking. *Haydocy Pontiac, Inc. v. Lee*, 250 N.E. 2d 898 (Ct. App. Ohio 1969). As a general rule, the minor may avoid both executory and executed contracts at any time during infancy and for a reasonable time after reaching majority provided that the contract has not been ratified. The right to disaffirm exists without regard to whether the contract is fair or favorable to the minor and exists without regard to whether the adult knew that the other party was a minor.

Ratification, which can be given after reaching majority, means to make the contract valid by approval. The minor's inaction or failure to disaffirm while enjoying the benefits of the contract for an unreasonable period of time, after reaching majority, will constitute ratification. Generally, no state permits ratification prior to the time the minor reaches majority.

Insane and Intoxicated Persons

Contracts with insane or mentally incompetent persons are generally voidable. A person who is temporarily insane may ratify or disaffirm previously made agreements upon becoming sane. However, under the majority view when the contract is made in good faith ignorance of the insanity with no advantage being taken of the incompetent person, the contract cannot be disaffirmed unless the parties can be restored to their original position. Generally, the same rules apply to persons so intoxicated that the person does not understand the nature of the transaction.

Principal/Agents, Partnerships, and Corporations

As the commercial economy expanded beyond simple person-to-person dealings, commercial necessity required that a person be able to act through others. From this necessity grew the concepts of principal/agent, partnerships, and corporations. Agency can perhaps be described as "let Jack do it." One party (a principal) employs another party (an agent) with whom third parties would deal. Third parties deal with the agent based on the assurance that they can look to the principal to fulfill the obligations. A more detailed discussion of agency and the forms of business entities is set forth in chapter 9.

Agency relationships are generally created through a manifestation by the principal to the agent that the agent can act on the principal's behalf and with some assent by the agent to so act. However, these elements of consent are often informal. Such assents need not be in writing except where statutes require that the transaction be in writing such as for the sale of land. The agent may also be a regular full-time employee of the principal. The agent's authority is classified as actual (real) or apparent (ostensible). Actual authority exists where the principal intentionally delegates power to the agent either expressly or by implication. Apparent authority is a manifestation of authority to a third party by the principal's actions permitting the agent to exercise power or holding out the agent as having authority. The edited opinion in the *Frank Sullivan case*, at the end of this chapter, illustrates a not unusual fact situation concerning an agent's authority.

In a partnership setting, the rules for determining the authority of a partner do not differ materially from the rules relating to the agent and principal. The courts agree that the relationship of partners to third persons is founded on the doctrine of mutual agency. The statutory statement of Section 9(1) of the Uniform Partnership Acts reads,

> Every partner is an agent of the partnership for the purpose of its business, and the act of every partner, including the execution in the partnership name of any instrument, for apparently carrying on in the usual way the business of the partnership of which he is a member binds the partnership, unless the partner acting in fact has no authority to act for the partnership in the particular matter, and the person with whom he is dealing has knowledge of the fact that he has no authority.

Every state has its own laws under which private business corporations are formed and operated. Generally, a corporation is formed by filing "articles of incorporation" (sometimes called "charter"). The corporation is owned by stockholders, and the corporation generally is governed in terms of broad policy by a board of directors elected by the stockholders. The board then employs officers (e.g., president, vice presidents, treasurer, secretary) who function pursuant to corporate bylaws that spell out in broad terms the duties of each officer. These officers, with the assistance of employees, conduct the day-to-day business of the corporation. The officers derive their authority from statutes, articles of incorporation, bylaws, and resolutions of the board of directors. The president, on behalf of the corporation, has at least implied authority to execute such contracts as are reasonably necessary and related to the business of the corporation. There is at least a rebuttable presumption that the president has the authority to perform any act within the scope of the corporation's business or anything the board could authorize the president to do.

WESTERN CONTRACTING CORP. v. SOONER CONSTRUCTION CO.
256 F. Supp. 163 (W.D. Okla. 1966)

Daughtery, District Judge. This is an action by the plaintiff, Western Contracting Corporation, against the defendant, Sooner Construction Company, for breach of an alleged subcontract between the parties on a runway project at Tinker Air Force Base, Oklahoma. There was no written subcontract signed by the parties. [T]he Court finds that Sooner orally quoted certain unit prices on asphalt paving to Western prior to Western submitting its bid on the project for the prime contract. Western was successful on its bid. Thereafter, Sooner confirmed its orally quoted prices to Western by a letter dated March 25, 1963.

Received after Bidding

Western Contracting Corporation
400 Benson Building
Sioux City, Iowa

Gentlemen:

   Congratulations on receiving the Tinker Field contract. I hope that it will prove to be a most successful job for you.

We would like very much to make a contract with you to do your asphalt paving work on this job. This letter is to confirm the prices which we quoted you at Fort Worth.

Item 9 Prime 48,150 gallons @ .19 Furnish, deliver and apply . . . no brooming or blotting.

Item 13. Hot mix surface 17,220 T.@ $8.32, less $.50 ton discount for payment by 10th of month;

Item 14. Asphalt (85-100) 215,350 gal. @ $.13

We will be happy to help you in any way possible on this contract. We would appreciate your contacting us when you establish your job office here. We hope the job will prove to be both pleasant and profitable and that we will have the opportunity of working with you.

Yours very truly,

/s/ Haskell Lemon

The oral quotes and the letter quotes were the same. Regarding the item of hot mix surface, the price quote of Sooner was 17,220 tons at $8.32 per ton less a $.50 per ton discount if payment is made by the 10th of the month.

During the period from March 25, and July 15, when Western opened an office at Tinker and the parties had various contacts, telephone calls, and discussions regarding the possibility of a subcontract, Western was obtaining asphalt quotes elsewhere. On July 15, at Tinker a meeting was held attended by a Mr. Hastie for Western, a Mr. Lemon for Sooner, a Mr. Pybas, a superintendent of Sooner, and a representative or representatives of the United States Corps of Engineers. At the meeting discussions were had regarding the specifications, equipment and rolling stock. Hastie testified that after this meeting, he for Western and Lemon for Sooner reached an oral agreement on the subcontract following which a form of subcontract, unsigned by Western, was forwarded by Hastie to Sooner for execution and returned to Western for execution by Western at its home office in Iowa. On the hot mix surface this written subcontract submitted by Hastie contained a price of $7.82 per ton thereon but did not provide for payment by the 10th of the month. Rather, it provided for partial payments to Sooner, less a retained percentage of 10% as Western was paid by the owner and final payment to Sooner upon complete performance of the subcontract within 45 days after final payment is received from the owner by Western. Lemon [Sooner] denied that an oral subcontract was agreed on July 15, with Hastie and denied that any discussions were even had whereby Sooner would agree to the $7.82 price with the retainage provision and final payment provision instead of payment for

the hot mix surface by the 10th of the month. Pybas, who testified that
he was with Lemon at all times going to, at and from the Tinker meeting on July 15, also denied any oral agreement on the subcontract or
any discussions about the discounted price of $7.82 being agreeable
without payment by the 10th of the month. Lemon testified that shortly
after receiving the written subcontract from Hastie he called Hastie on
the phone several times and objected to the lower price of $7.82 per ton
without payment being provided for by the 10th of the month in accordance with his quoted terms. Hastie acknowledged several telephone
conversations after July 15, 1963, with Lemon regarding the retainage
and that Hastie suggested in one of these conversations a reduction of
the retainage to only 50% of the work. Hastie further testified that
Lemon never gave him an answer to this suggestion.

   In late September, 1963, certain developments took place. Sooner
sent a signed subcontract to Western to which it attached certain
amendments, six in number, one of which called for full payment for
each of the three phases of the work to be done by Sooner within 45
days of completion by Sooner of each phase. [These amendments were]
in lieu of Sooner's requirement in its written confirmation of payment
by the 10th of the month and Western's requirement in the written
subcontract it prepared and submitted of the 10% retainage and the
final payment in 45 days. Western wrote Sooner a letter advising
Sooner that it was delinquent in the performance of its subcontract and
that if Sooner did not correct this default in performance within five
days Western would exercise its rights under the subcontract. Then on
September 29 or 30, a meeting was held in Oklahoma City which
brought a Mr. Shaller down from Iowa for Western. Shaller as manager
of heavy construction for Western was over Hastie. At this meeting, the
six amendments were discussed one by one and Shaller disapproved the
amendment about full payment in 45 days following each phase of completion as well as two other amendments and in his own hand wrote
"out" opposite each of the three amendments so disapproved. In the language of Shaller, finally at this meeting "things were terminated."
Under date of October 2, Western made a subcontract with
Metropolitan Paving Company for larger unit prices as to all items (the
price for hot mix surface—the largest item—was $8.53 per ton) and
sues herein for the difference amounting to $16,957.08 plus interest,
overhead and profit and other expenses.

   "In order that a counter-offer and acceptance thereof may result in
a binding contract, the acceptance must be absolute, unconditional, and
identical with the terms of the counteroffer."

   [In *Fry v. Foster* the court stated] . . . "Where parties to an agreement make its reduction to writing and signing a condition precedent to
its completion, it will not be a contract until this is done, and this is

true although all the terms of the contract have been agreed upon. But, where parties have assented to all the terms of the contract, and they are fully understood in the same way by each of them, the mere reference in conjunction therewith to a future contract in writing will not negative the existence of a present contract."

The Court is of the opinion and finds and concludes that Hastie and Lemon did not on July 15, reach an oral agreement on the subcontract or discuss and settle the price differential on the hot mix surface with reference to the two alternative prices quoted by Sooner and the effect of the method of payment on the same. It is believed that when Western prepared the subcontract shortly after July 15, 1963, it sought to take advantage of the lower quoted price without meeting the condition attached to the same regarding payment. To this, Sooner promptly objected and Sooner did not sign and return the subcontract as requested by Western. Hastie admits that several telephone conversations immediately followed his mailing the subcontract he prepared, these telephone conversations coming from Lemon and that the subject matter of the calls had to do with the retainage provision of the submitted subcontract as it affected the price of the hot mix surface.

The Court, therefore, finds and concludes from the evidence that Sooner quoted a price of $7.82 a ton on hot mix surface provided payment was received for the same by the 10th of the month, otherwise the price would be $8.32 a ton; that the weight of the evidence indicates that this alternative quote and payment condition of Sooner was not changed or discussed on July 15; that Western in submitting a subcontract to Sooner set out the $7.82 per ton price but did not meet the payment condition attached to the same; that Sooner immediately and admittedly by several telephone conversations objected to this feature of the subcontract as submitted; that while a period of several weeks lapsed before Sooner submitted its subcontract with amendments such lapse of time transpired in the face of and after objections were made by Sooner to the subcontract submitted by Western and particularly with reference to the use of the lower quoted price on hot mix surface without complying with the requested method of payment in the use of such price; that at the meeting on September 30, Western would not agree to the originally quoted alternative price of Sooner or its modification as later proposed by Sooner and terminated the matter . . . . The Court is of the opinion that the parties never reached a meeting of the minds on the price and method of payment for the hot mix surface, the principle item involved.

Plaintiff is, therefore, not entitled to the judgment it seeks.

FRANK SULLIVAN CO. v. MIDWEST SHEET METAL WORKS
335 F.2d 433 (8th Cir. 1964)

Blackmun, Circuit Judge. Midwest Sheet Metal Works [plaintiff], a Minnesota partnership, instituted this diversity suit against Frank Sullivan Company [defendant], a Boston contractor, to recover damages for breach of contract. The jury returned a verdict in favor of Midwest for $85,000.

[This] controversy arises out of the project for the extension and remodeling of the United States Post Office and Customs House at Saint Paul, Minnesota.

[The two primary issues on appeal are]: (1) The identity of Midwest's Exhibit 5 as the agreement between the parties, and its effectiveness as a contract [which was admitted in evidence over the defendant's objection]. (2) The authority of Sullivan's agent to sign the agreement. The prime contractor on the project was Electronic & Missile Facilities. Inc., of New York City (EMF). On December 4, 1961, EMF and Sullivan executed a lengthy and detailed subcontract whereby Sullivan undertook all plumbing, heating apparatus, air conditioning and ventilation work on the job for an agreed price of $1,650,000. This contract was executed on behalf of Sullivan by Francis J. Sullivan (Frank), as its president.

In the fall of 1961, before the formal execution of the agreement between EMF and Sullivan, Frank had contact with Michael J. Elnicky, the dominant partner of Midwest, about Midwest taking on the sheet metal and air conditioning portion of Sullivan's subcontract. Sullivan had even invited a quotation from Midwest for this work. On November 1, Midwest quoted a figure in excess of a million dollars. Frank testified he telephoned and told Elnicky that this bid was about $200,000 too high. Elnicky indicated he might reduce his price somewhat but could not approach Sullivan's suggested figure. Frank testified that he then told Elnicky, "Well, look, we have got a fellow going out there, and I will show you that these are not prices we dreamed up, these are prices that we used in making our bid, and we got them confirmed by letters of reputable people."

Near the close of 1961, EMF told Sullivan that it was imperative that the work in Saint Paul be started. Sullivan promised EMF that it would get a superintendent, a foreman, and men and material on the job by the end of January. Sullivan sent John Sullivan (Jack) from Boston to Saint Paul in early January. On this trip he conferred with EMF's superintendent on the project. Jack was back in Minnesota later in the same month with Byers who was to be Sullivan's general super-intendent on the job. Before he left Boston on this second trip, Jack had been instructed by Frank to look over the labor situation in the area, to check in with prospective subcontractors, to see Elnicky and give him quotations which "will back up the reduction of his bid," and to "get the job started." Frank gave Jack the job estimates which had been prepared by Sullivan but he was not given and had not seen the prime contract.

Upon their arrival Byers called upon local union business agents, purchased material and tools, received other equipment from Boston, and placed four steamfitters on the job.

On Monday, January 22, Jack came to Midwest's office. This was the first time Elnicky saw him. Elnicky and some of his employees tes-tified that Jack told him on this visit that he was "part of the [Sullivan] organization" and had "a piece of it." Jack denied that he made any such statement. Elnicky conceded that he made no attempt to check Jack's authority with the Sullivan home office and that he did not ask for written evidence of it.

Elnicky also testified that Jack early in this first meeting sug-gested that Midwest price off "the whole works"; in any event, Elnicky indicated that he was interested in taking over the entire Sullivan job. Jack did not object and said that he was trying to get a copy of the prime contract for him. The two men met again on Tuesday when Jack permitted Elnicky and his people to review the bids Sullivan had received. By Wednesday, Jack obtained a copy of the prime contract from EMF's office on the job. He gave it to Elnicky who kept it over-night. Meanwhile, Jack talked with other prospective subcontractors. Elnicky and Jack met further, and sometimes socially, during the same week. Jack told him that Sullivan would have to have a minimum of $100,000 if Elnicky took over. On Thursday Jack told Frank by telephone of the discussions he was having with Midwest. As to this conversation Frank testified that he told Jack that this could not be done, that EMF wanted Sullivan on the job, and that Jack should "pick up what you got and come home." Midwest finished its estimates on Saturday, January 27. That evening Jack was at Elnicky's home with two of the latter's men. They discussed costs and what Elnicky might offer to do the job but no conclusion was reached.

Early in the afternoon of the next day, Sunday, Elnicky came to Jack's hotel room. Jack was planning to return to Boston. Elnicky ar-rived with a fifth of Scotch. The men were together for three and one-half hours, discussing the job and drinking the entire fifth. Byers was present but left for a time to get the copy of the prime contract which had been left elsewhere. Elnicky made an offer of $1,550,000 to per-form the work. This was discussed as was the question of what to do

with the equipment and materials which Sullivan already had on the
job. Jack then started to write something out. Elnicky dictated part of
it. Several drafts were made. Later Jack dictated a draft to a hotel
typist and he and Elnicky signed it. Elnicky then took Jack to the airport. The typed draft is Midwest's Exhibit 5 and is the document in
controversy. It was admitted in evidence over Sullivan's objection.

The testimony as to the execution of the exhibit is in sharp conflict. Jack testified that he told Elnicky that this agreement was subject
to approval by Frank and EMF, that there was no sense in working out
other details until this was done, that it was his intention that Exhibit
5 be merely a proposal by Elnicky to Sullivan, that he did not intend
thereby to turn the Sullivan contract over to Midwest. Byers testified
that Jack told Elnicky that it had to be approved by Frank and EMF;
that Elnicky acknowledged this; and that he, Byers, had expressed a
hope that Frank would approve it so he "could go home." Elnicky flatly
denied that Jack had said Exhibit 5 was subject to approval by Frank
and EMF.

Exhibit No. 5

Hotel Saint Paul
St. Paul, Minnesota
January 28, 1962

The Midwest Sheet Metal Co.
340 Taft St.
Minneapolis, Minnesota

The Midwest Sheet Metal Company of 340 Taft St., Minneapolis, Minnesota, agrees to take over Frank Sullivan Company's contract with Electronic
Missile Facilities,Inc.,in the amount of One Million Five Hundred Fifty and
00/100 Dollars ($1,550,000).

The cost of the bond will be paid for by the Frank Sullivan Company.

The Midwest Sheet Metal Company will man the above mentioned project
on January 29, 1962, to show good faith regarding this contract.

The above agreement is made between

/s/ Midwest Sheet Metal
M.J. Elnicky
Frank Sullivan Co.
John Sullivan

The next day, Monday the 29th, Jack called Byers from Boston to
see if Elnicky had someone on the job as the writing provided. Byers
called Elnicky who told him he would have someone there on Tuesday.
On Tuesday, Jack and Elnicky conferred by telephone. Byers then sent

Elnicky's men away from the job. On the same day, January 30th, Jack, signing on behalf of the Sullivan Company, wrote Midwest that "the agreement made on January 28th, 1962, between the Midwest Sheet Metal Company and Frank Sullivan Company is hereby canceled" and that they would try to arrange for Midwest to quote on the job's ventilating, air conditioning, and refrigeration. Jack testified that he wrote this letter without discussion with Frank. Midwest's receipt of that letter led to the present suit.

Jack Sullivan's status is obviously of vital importance. He was 29 years of age. He was a high school graduate and had had one year of "night college." He held a plumber's union card and a journeyman plumber's license. He had worked for Sullivan for ten years [starting] as a plumber and [later promoted] to a job superintendent. [Subsequently] he was assigned as an "outside superintendent." In this capacity he traveled to various jobs, examined labor and material situations and reported back to Sullivan. He did no hiring or firing. He did not order materials. He did make recommendations. He was not an officer, director, or shareholder of Sullivan. He was not related to Frank. He had nothing to do with obtaining or negotiating subcontracts. He had had no experience in sheet metal or air conditioning. He had done no estimating. He had been given no specific authority to sign any contract for Sullivan or to assign Sullivan's subcontract with EMF.

Elnicky was about 52. He had been in sheet metal and similar work for many years and had run his own business [for approximately 15 years].

Sullivan's basic argument here is that Exhibit 5 . . . is not sufficiently clear and definite to be valid and to constitute an enforceable contract.

[If]. . . substantial terms are left open and subject to further agreement, which is never reached, there is not only no complete agreement but no contract at all. The next inquiry is whether, there being an agreement which is asserted to be a contract, it is complete. It is not unless it is in all essential terms definite and certain or capable of being made so by the aid of competent evidence and permissible interpretation. If as a contract it be incomplete, a court can no more complete it for the parties than it could make it for them in the beginning."

There is no contract where there is no mutual and final assent to all the essential terms of a bargain. If an alleged contract is so uncertain as to any of its essential terms that it cannot be carried into effect without new and additional stipulations between the parties, it is not a valid agreement. And "where substantial and necessary terms are specifically left open for future negotiation; the purported contract is fatally defective." [But] . . . a proper administration of justice does not permit an over-zealous quest for subtle ambiguity to destroy the intent of the parties when the court, despite some incompleteness and imperfection of expression, can reasonably find that intent by applying the words used, with all their reasonable implications, to the subject matter as the parties themselves . . . ." This court is reluctant to invoke the principle that indefiniteness prevents the creation of a contract where a just result, consistent with a reasonably expressed intent of the parties, can be reached by upholding the agreement.

While Exhibit 5 may have been born with somewhat of an alcoholic background, this fact alone does not make it any less a contract. Evidently both Elnicky and Jack desired or were content to negotiate in that kind of atmosphere. It is not now claimed that there was duress or that Elnicky took advantage of Jack through alcohol.

Exhibit 5 on its face is clear and complete. It succinctly states that Midwest "agrees to take over" Sullivan's contract with EMF. It states that figure at which it does so. It leaves a $100,000 gross profit margin, less the cost of the bond, for Sullivan. It calls for immediate manning of the job Midwest. It contains no ambiguity and Sullivan so concedes.

Although, as Sullivan observes, Exhibit 5 consists of but one page, as contrasted with the many pages of the subcontract between EMF and Sullivan, this difference is understandable. The document could have been more formal but it was formulated by two laymen and it in effect incorporated the detailed EMF subcontract by reference. By doing so it achieved certainty as to underlying details. Sullivan itself was originally content in this respect or it would not have made the subcontract with EMF.

The discrepancy between the words and the figures of Exhibit 5 is not fatal. It is true that the normal rule is that, where there is a discrepancy of this kind, the words and not the figures control . . . . This usually does not invalidate a contract. Each knew that the proper amount was $1,550,000. No one was misled.

Jack's own letter of January 30, is almost persuasive in itself. It refers to "the agreement" of January 28 "between Midwest . . . and . . . Sullivan" and reaches out to cancel it.

[The issue of Jack's authority]. Midwest does not contend that the record supports a finding that Jack possessed actual authority to act on behalf of Sullivan. The issue is one of apparent authority.

Jack certainly assumed the mantle and the posture of responsibility and authority. There is evidence to the effect that he professed an ownership interest in Sullivan, possessed and produced the bids and the cost estimates the company had assembled, permitted Elnicky to review them, obtained a copy of the EMF-Sullivan subcontract for Elnicky, mentioned to others than Elnicky his interest in contracting out the entire Sullivan portion of the job, demonstrated a permissive attitude toward Midwest's interest in taking on the full subcontract, was in contact with Elnicky's performance bond man and asked him to confirm his comments by letter, accepted Elnicky's entertainment favors, bargained continuously for a week, and even wrote the letter of cancellation.

But apparent authority must be founded on something more than the conduct and statements of the agent himself. Liability can be imposed upon a principal only "for that appearance of authority caused by himself." 2 Williston on Contracts (3d Ed. 1960), Section 277A pp. 222-224 . . . Of course, a degree of reasonableness and diligence is required of one who deals with the agent . . . .

We find in this record adequate support for the submission of the issue of apparent authority to the jury. Accepting the evidence, as we must, in the light most favorable to the prevailing plaintiff, we have, apart from and in addition to Jack's own acts and statements, all the following: (1) Sullivan sublet part of every job it had, for there were certain types of work (air conditioning and sheet metal, for example) which it never performed; (2) Sullivan sent Jack to Saint Paul to get the job started; (3) Sullivan instructed Jack to get in touch with area people in the construction industry and to obtain subcontract offers; (4) Sullivan placed Jack in possession of the breakdown of costs it had prepared and of the bids it had received; (5) Frank told Elnicky that he had a man going out to Minnesota; (6) Jack possessed the Sullivan name; (7) Jack was the highest person in authority in Sullivan's employ on the job; (8) so far as the Saint Paul job was concerned, Byers followed Jack's instructions; (9) Jack, while he was in Saint Paul the week of January 22, was in telephone communication with Boston; and, specifically, talked with Frank; and (10) Frank knew that Jack was discussing with Elnicky a complete takeover by Midwest and yet did nothing to disavow his status to Elnicky. And, for what it is worth, Jack was still with the Sullivan Company at the time of the trial.

Of course, there is an opposing factual argument, namely, that Sullivan had given Jack no instructions to turn over the entire job; that Jack was not supplied with a copy of the EMF-Sullivan subcontract and had to obtain it from the EMF man on the job, that the preliminary conversations between Sullivan and Elnicky had only to do with a limited area of work; that Elnicky knew who Frank was and was in communication with him; and that Elnicky did not inquire of Jack or of Frank as to Jack's authority.

Sullivan places great emphasis on Elnicky's failure to make inquiry as to Jack's authority, and it urges the important nature of the contract as demonstrated by the amount involved and the time required for its performance. This argument, however, cuts both ways. If the job was so large and so important, a jury might properly infer that Sullivan's top man on the project was there with workable authority.

[In an earlier case,] Chief Justice Knutson, in speaking for the court, said, "Apparent authority exists by virtue of conduct on the part of the principal which warrants a finding that a third party, acting in good faith, was justified in relying on the assumption that the agent had authority to act."

The situation here strikes us as one where, as the negotiations developed, Jack sensed the opportunity to bring his company out of the project at a convenient profit of $100,000, without additional cost or participation; and further sensed that this would be a feather in his cap if he could bring it about. Whether [defendant Frank] Sullivan sensed this, desired it, and encouraged it, we shall, of course, never know with positive assurance. But the record supports just such an inference by the jury. We are not at liberty to overturn that body's conclusion. Affirmed [for plaintiff Midwest].

DISCUSSION QUESTIONS

1. What are the essential elements of a valid contract? Describe a voidable contract, and illustrate the essential elements that would most commonly be at issue?

2. Discuss the conditions under which an offer terminates.

3. If the offer contains a clause that "this offer will remain open for 60 days," can the offeror legally or ethically terminate the offer before the end of the 60-day period?

4. Describe the various means by which an offer may be accepted. Does the Western Contracting case suggest the criteria to be used by the court to distinguish between a counteroffer and negotiations following a bid quote?

5. Illustrate a situation in which a remedy in quasi-contract would be sought. Does the fact that the party seeking to recover in quasi-contract originally performed some of the work under a contract affect the outcome of an action to recover under quasi-contract?

6. In the Western Contracting case, what does the court conclude about when a contract comes in existence in situations where the parties reach an oral agreement and intend to reduce that agreement to writing?

7. In the Frank Sullivan case, how and with what facts does the court consider the issue of Jack Sullivan's authority? If the court decided that Jack Sullivan did not have apparent authority, what would have been the resulting positions of the parties?

4

CONTRACT INTERPRETATION, STATUTE OF FRAUDS, AND ASSIGNMENT

4.1 IS MEETING OF THE MINDS NECESSARY?

There is a common expression that a valid contract requires a "meeting of the minds." Williston, the outstanding scholar of contract law, in tracing the formulation of this short-hand phrase, explained that the democratic West, as opposed to the Communist nations, where terms of the agreement are imposed on the parties, had always placed great emphasis on the ego (i.e., the individual will in the formation of contract); hence, the development of the phrase "meeting of the minds." Although that phrase is useful in describing some of the aspects of contract formation, the law does not literally require a meeting of the minds for the validity of a contract. It is not a function of the law to determine whether there has been an actual meeting of the minds, it is a psychological function and not a legal function. Since such a probing of the minds of the parties would be totally unacceptable in the commercial world, the rule has been established that rather than attempt to ascertain the subjective intention of the parties, the law looks to their objective conduct. So, as discussed in chapter 3, an offer made in jest or with a secret reservation, but which otherwise appears valid, creates a power of acceptance in the offeree unless a reasonable person would have understood the offer to be in jest.

At times, courts must determine under some circumstances whether one or both of the parties genuinely intended to enter into a contract. Those questions are relevant inquiries in situations of alleged fraud, duress, mistake, undue influence, and unconscionability in the formation of the contract.

Mistake

Contracts are frequently attacked on the basis of mistake in the formation process of the agreement. Underlying every contract are certain assumptions made by the parties but not expressed. One possibility is a mistake involving an underlying assumption, which is not true, thus making the contract undesirable. For example, the contracting parties generally assume that the subsurface conditions to be encountered will not vary significantly from those expected by the design professional and the contractor. Although some deviation may be expected, drastic deviation that would have a tremendous effect on the contractor's performance may be beyond the mutual assumptions.

Historically, the law has considered mistakes as being of two kinds: unilateral and mutual. Generally, if only one party makes a mistake concerning a fact pertaining to the contract, that party is bound by the contract, and the unilateral mistake is not grounds for rescission or an excuse for nonperformance. However, some courts, in allowing relief involving unilateral mistake, have undertaken to balance the advantages and disadvantages to both parties if the contract were declared void. In making that determination, the court considers, first, whether enforcement of the contract would be oppressive to the party who made the mistake and, second, whether voiding it would result in a substantial detriment to the other party.

On the other hand, a mutual mistake as to any material aspect of the contract's subject matter usually provides grounds for either party to rescind or reform the contract since these mistakes strike at the element of mutuality of agreement. *Gevyn Construction Corp. v. United States*, 357 F. Supp. 18 (S.D.N.Y. 1972), involved a construction contract with the U.S. Post Office department in which a Michigan official sent a letter to the Post Office department indicating that the contractor could tap into a designated storm sewer of the state. Although the contractor relied on this in making a bid, the state subsequently refused permission to make the tap, and the contractor was forced to connect to the drain at a more distant point. The false assumption of both contracting parties was the basis of court relief.

Misrepresentation and Fraud

An element of a valid contract is that the agreement must have been entered into voluntarily, fairly, and honestly. Misrepresentation is the creation in the mind of another person of an impression not in accordance with the actual facts. A misrepresentation involving a material fact on which the other party was entitled to rely prevents formation of a contract and justifies avoiding the contractual promise by the party that was misled. "Materiality" means a significant and contributing factor in the decision to enter the contract. Generally, the misrepresentation need not be intentional and whether the party making the misrepresentation knows it to be false (innocent misrepresentation) does not affect the voidability of the contract. However, the party to whom the misrepresentation was made must neither know nor be in a position, in view of experience or all of the other facts, to have discovered the falsity. Less than a full disclosure may constitute a misrepresentation. In *Ikeda v. Curtis*, 261 P.2d 684 (Wash. 1951), a seller of a hotel was held to have made a misrepresentation by describing the source of income as from permanent and transient guests but failed to mention that the greater portion of the income was from single-night stop overs by sailors using the hotel as a brothel.

Fraud, as distinguished from misrepresentation, involves an intention to mislead amounting to what the law terms "scienter." The elements of fraud are generally described as (1) a false statement or the concealment of a material fact, (2) knowledge of the falsity of the statement or conduct taken with utter recklessness for the truth, (3) justifiable reliance on the false statement by the other party, and (4) injury to the other party as a result of the reliance. Where the misrepresentation was made intentionally amounting to a fraud, the injured party without regard to the materiality of the fraudulent misrepresentation may elect to rescind the contract or affirm the contract and bring a tort action for deceit.

Duress and Undue Influence

Contracts made under duress or undue influence are voidable. The lack of free will strikes at the concept of mutual assent, which is an underpinning in the doctrine of contracts. Early Common Law duress was physical, such as threatening bodily harm to the other party; but, in the modern context, duress has been extended to economic duress or business compulsion where one party exerts excessive pressure beyond permissible bargaining.

The RESTATEMENT of Contracts, Section 177, refers to factors including "the unfairness of the resulting bargain, the unavailability of independent advice, and the susceptibility of the person persuaded in which the party consents because there is no real choice." In *Austin Instrument, Ins. v. Loral Corp.*, 324 N.Y.S.2d 22, 272 N.E.2d 533 (1973), the prime contractor under a federal procurement contract contended that the supplier forced the contractor to modify the prices paid to the supplier under threat that the supplier would stop deliveries. Contractor, having no other available supplier, was faced with substantial liquidated damages for delay since the government might terminate the contract for failure to deliver and jeopardize the chance for additional contracts. The court concluded that this pressure constituted undue economic duress and relieved the prime contractor from paying the modified contract prices.

Unconscionable Contracts

A doctrine is now emerging that looks at conduct short of misrepresentation to hold that contracts must be fair and formed in good faith. An explanatory note to Uniform Commercial Code (UCC) Section 2-302 clarifies the meaning of "unconscionable" by describing the basic test as "whether in the light of the general commercial background and the commercial needs of the particular trade or case, the clauses involved are so one-sided as to be unconscionable [oppressive or grossly unfair] under the circumstances existing at the time of making the contract." The court will hear evidence on these questions to determine whether one of the parties was so oppressed that the contract will not be enforced.

In *Henningsen v. Bloomfield Motors, Inc.*, 161 A.2d 69 (N.J. 1960), plaintiff, in purchasing a new car, signed a sales contract stating that "the dealer and manufacturer made no express or implied warranty except that defective parts would be replaced within 90 days of purchase." In that case, the plaintiff purchaser was permitted to recover for personal injuries suffered as a result of a faulty steering mechanism. The court noted that automobile manufacturers are few in number, that they are in a strong bargaining position with ultimate purchasers, and that the dealer had no authority to change this clause in the contract with the ultimate purchaser. While denying that refusal to enforce the warranty limitation was a restriction on freedom of contract to allocate risks, the court expressed the doctrine of unconscionability as a "judicial limitation to prevent oppression and unfair surprise."

4.2 SUBJECT MATTER

A valid contract requires that agreements have a lawful purpose and object and that those that fail to meet this standard are unenforceable. If the contract is executory, the court will not enforce the unperformed duties. If the contract is executed, the court will not order rescission. In effect, the court leaves the parties where it finds them refusing to intercede to rectify the transaction since the purpose is to deter illegal bargains. A contract may be declared illegal if it is specifically prohibited by statute such as contracts improperly intended to influence government by bribes, to fix prices, or to charge usurious rates of interest. In addition, the court may hold that the contract is illegal if it is contrary to public policy. For example, an exculpatory clause, such as a provision in a contract relieving a party of liability for negligent conduct, is not favored by the law and frequently is declared illegal. The rationale behind these decisions is based on the unconscionable demands or the unequal bargaining power of the parties. In *Van Hosen v. Bankers Trust Company*, 200 N.W.2d 504 (Iowa 1972), the defendant bank unsuccessfully contended that the plaintiff, who retired after 33 years of service with defendant after advancing from messenger to vice president, had forfeited his rights to receive annuity payments under defendant's pension plan because plaintiff accepted employment (violating a non competition clause in the pension plan) with the commercial loan department of a competing Des Moines bank. The court noted that the concept of employee rights and the stature of employee fringe benefits in the industrial world had undergone a substantial change over the years. While recognizing that an employee pension program has legitimate business interests in seeking to encourage loyal and productive career service on the part of the participants, nevertheless, there are counterbalancing factors that include the humanitarian purpose of the pension that has become part of the socioeconomic community expectation. The court concluded that to deprive the plaintiff of the right to work in the field of his experience amounted to a forfeiture imposing an unjust and uncivic penalty on the plaintiff disproportionate to the benefit to the defendants.

4.3 THIRD-PARTY BENEFICIARIES

Under English Common Law, only parties to a contract had enforceable rights under a contract based on the theory that privity of contract was necessary to bring an action and recover. *Price v. Easton*, 110 Eng. Rep. 518 (K.B. 1833). The courts, using the strict logic of extreme individualist psychology, reasoned that a person who did not give manifest assent or promise could not obtain a contractual right in an agreement of other parties. In a more modern setting, the California Appellate Court in *G & P Electric Co. v. Dumont Construction Co.*, 194 Cal. App. 2d 868 (1961), held that the owner of a building could not claim damages from the electrical subcontractor for failure to comply with the requirements of the subcontract between the prime contractor and the electrical subcontractor. The court held that since there was no contractual relationship between the owner and the electrical subcontractor, the owner lacked the privity of contract on which the litigation could be based.

Most states recognize a doctrine known as third-party beneficiaries in contracts that confer benefits, either directly or indirectly, on third-parties. The effect of a third-party contract is to give a legal right to a person who is not a party to the contract. Traditionally, the courts have classified third-party beneficiaries in three categories: (1) creditor-beneficiaries, who have enforceable rights; (2) donee-beneficiaries, who have enforceable rights; or (3) incidental-beneficiaries, who have no enforceable rights. Under the RESTATEMENT (Second) Contracts Section 133, the terms "creditor" and "donee" are eliminated and a new term, "intended" beneficiary, is substituted.

A third-party beneficiary is not entitled to enforce a contract unless it is established that the parties actually intended that the third-party be benefited by the contract. The intent to benefit the third-party must appear from the terms of the contract, but such intent is easily inferred in creditor-beneficiary situations. The third-party need not be named as an individual in the contract, and the fact that the actual contracting party could also sue to enforce the agreement will not bar a suit by the beneficiary. In most states, a contract made for the express purpose of benefiting a third-party may not be rescinded without the consent of the beneficiary after its terms have been accepted by the beneficiary. The beneficiary has a vested interest in the agreement from the moment the contract is made and accepted.

Creditor-Beneficiary

Where a contract between X and Y provides that Y will discharge a debt that X owes to Z, the effect is that Z is a creditor-beneficiary. Likewise, beneficiaries of a life insurance contract are third-party beneficiaries, as are owners of buildings under construction who may not be parties to a performance bond obtained by the contractor.

Donee-Beneficiary

The characteristic features of a donee-beneficiary contract and a creditor-beneficiary contract are similar except that the purpose of the donee-beneficiary contract is to make a gift to a third person. However, the donee-beneficiary contract differs from the creditor-beneficiary contract in that the creditor-beneficiary may sue either of the parties to the contract while the donee-beneficiary can obtain a remedy only from the person who promised to perform for the benefit of the donee.

Incidental-Beneficiary

A contract entered into primarily for the benefit of the contracting parties, which may indirectly or incidentally benefit some third person, is an incidental-beneficiary contract. The contracting parties cannot be sued by such third person. For example, the breach of a contract between X, an owner, and Y, a contractor, to erect a new building that would incidentally benefit Z, an adjacent property owner, does not give Z, the incidental-beneficiary, any right to sue.

4.4 ASSIGNMENT OF CONTRACTS

Right to Assign

In early Common Law, contracts were not assignable, since contracts were considered purely personal rights that could not be transferred. As business practices developed the use of the property concept to include intangibles (e.g., bank accounts and accounts receivable), the legal doctrine recognized the necessity of assignments as an aid to commerce.

An assignment of a contract is the transfer to another person of the assignor's rights to performance under the contract. For example, X, a steel company, sells girders to Y, a contractor, for $10,000, payable 15 days after delivery of the steel to the job site. After delivery of the steel, X transfers the right to receive payment to Z, X's bank, and notifies Y of this transfer. X ("assignor") has made an assignment of the contract to Z ("the assignee"), and Y ("obligor") must pay the $10,000 purchase price to Z.

Today, most contracts are assignable unless the nature of the contract or its terms show that the intention of the parties was to make the contract personal and unassignable. If the contract provides that any assignment shall be void, the words are given their obvious meaning, and no effective assignment can be made. On the other hand, a general provision against assignment is usually interpreted to prohibit the assignment of duties but not of rights. So, in *Norton v. Whitehead*, 24 P. 154 (Cal. 1890), where a building contract provided that "this contract shall not be assigned" and the builder after performance of the work assigned the right to receive payment, the assignment was held to be valid. The court reasoned that the construction work was not assignable, but the right to receive payment was relatively unimportant and permitted assignment of payments where the clause was general.

It is frequently stated as a truism that "rights may be assigned but liabilities may not be assigned." More specifically, it is stated, unless assignment is prohibited by the contract or a statute, that most contracts are assignable except where the assignment would (1) materially change the duty of the other party (for example, if X has contracted for Y to design a building, Y's duties under the contract cannot be assigned to Z since X is entitled to the creativity and competency of Y, and Z may be quite different); (2) materially increase the burden of risk imposed by the contract (for example, if X contracts to supply Y with Y's requirements of concrete, Y cannot assign the contract to Z who has different supply requirements); or (3) materially impair the other party's chance of obtaining return performance (for example, if X agrees to finance Y's building project, Y cannot assign that contract to Z who takes over the building project because the chance of having the loan repaid may be less owing to Z's lower credit rating).

Assignment: Effect on Rights of Assignor

The assignment of a contract having both rights and duties consists of really two phases: (1) the delegation of the power to perform the assignor's contractual obligations assuming that the assignment is not otherwise prohibited and (2) assignment of rights of the assignor under the contract. With respect to the obligation to perform, the assignment does not relieve the assignor of the obligations to the other contracting party. The assignor is still liable if the assignee fails to perform as agreed, in which case the other contracting party would have a cause of action against the assignor.

Assignment: Effect on Rights of Assignee

The assignee of a contract can obtain no greater rights in the contract than the assignor had, since the assignment is essentially a sale of the assignor's contract rights. The assignee takes the contract subject to all defenses that the promisor had against the assignor on the contract. The assignment of the contract transfers the benefits subject to the burdens. The rule is usually stated as the "assignee stands in the shoes of the assignor." The assignee, as the real party in interest, may sue the other contracting party to enforce the rights assigned.

Under a general assignment of an executory contract, the rule of RESTATEMENT (Contracts) and the case law hold that an assignee, in the absence of circumstances showing a contrary intention, becomes the delegatee of the assignor's duties and with an implied promise to the assignor that assignee will perform such duties. The fact that the assignee takes the assignment subject to defenses does not mean that the assignee has no recourse against the assignor. In making the assignment for value, the assignor makes implied warranties that the parties had capacity to contract, and that the claim or right is valid and not void because of illegality and that it has not been discharged.

To protect the rights acquired by the assignment, notice of the assignment should be given to the other contracting party. Where the assignor makes more than one assignment, the "American rule" provides that the first assignee has the better right while the "English rule" provides that the assignee who first gives notice of the assignment has the better right.

4.5 STATUTE OF FRAUDS

Although oral agreements are generally enforceable, the Statute of Frauds provides that certain kinds of agreements are unenforceable unless evidenced by a legally sufficient memorandum. In England during the reign of Charles II, the courts were faced with many cases of fraud concerning oral agreements, and instances of perjury were frequent as parties attempted to prove the existence of contracts. In 1677, the English Parliament passed the Statute of Frauds to protect litigants from dishonest claims and to relieve the courts of the burden of hearing claims of questionable merit. The Statute of Frauds has been adopted in some form in all U.S. jurisdictions, and, generally, four kinds of contracts are within the statute, namely,

(a) contracts which create or transfer any interest in land although most states exempt oral leases for less than one year;

(b) contracts not to be performed within one year [where the contract is not explicit about the time within which performance is expected to be complete, courts usually supply a liberal construction and hold that the test is the "shortest possible" and not the "probable" time required for performance so that a contract that might be performed in less than a year is not within the statute];

(c) contracts for the sale of goods having a price of $500 or more; and

(d) contracts to answer for the debt or default of another.

Cases can be removed from the Statute of Frauds by certain actions or conditions. Although some states require a written contract, the Statute of Frauds in most states provides that a memorandum is sufficient. At any time prior to filing the law suit, either the memorandum or the contract may be made and in some cases a memorandum may consist of several documents showing that they should be taken together as evidence of a single contract. The memorandum or contract must include the names of the parties to the contract, contain the material terms of the understanding, and describe the subject matter of the agreement with reasonable certainty. In addition, the memorandum or contract must be signed either by an authorized party or by the party against whom enforcement of the contract is to be sought. An oral variation of a written contract not evidenced by a sufficient writing or memorandum is not enforceable, although the original contract that is evidenced by sufficient writing would be enforceable.

An oral contract within the Statute of Frauds is not void or voidable but merely unenforceable, since the court will not aid in its enforcement. A party who has performed such a contract may not recover on the contract but, in certain cases, may recover in quasi-contract for the value of the benefits conferred on the other party.

4.6 STATUTE OF LIMITATIONS

The Statute of Limitations prescribes a time limit within which a law suit must be started after a cause of action arises. Different time limits are prescribed for various causes of action, such as the recovery of land, breach of contract, and negligent injury.

Since time limitations differ based on the theory of the claim, a host of questions have arisen under these statutes concerning whether a particular action brought is based on a claim of product liability, breach of contract, or mere negligence. In addition, other questions have arisen such as whether the statute starts running from (1) the date of the completion of the project, (2) discovery of a defect, or (3) injury based on the defect.

Product liability cases provide illustrations of the quandary presented. UCC Section 2-725 provides that actions for breach of contract must be brought within 4 years after the right to sue accrued. Generally, this right to sue would accrue at the time the breach of contract occurred; but, under a breach of warranty theory, the right to sue arguably begins at the time of delivery of the product. In Georgia, the legislature has declared that no product liability suit may be filed more than 10 years following the first sale or use of the product. On the other hand, the Oregon statute provides that strict liability suits must be commenced within 2 years following the date of injury; however, no suit may be begun more than 8 years following the first sale of the product.

Recognizing that engineering professionals work on projects involving astronomical potential liability while their fees represent only a small percent of the project cost, architects and engineers, through their professional organizations, have urged state legislatures to adopt laws limiting the time in which suits can be filed based on malpractice. To shorten the time period in which claims from owners may be filed, AIA Document B141 contains a stipulation that the Statute of Limitations "will begin to run no later than the date of substantial completion." While useful between the contracting parties, this contract provision has no effect on claims by third parties.

4.7 CONSTRUCTION AND INTERPRETATION OF CONTRACTS

General

Courts are frequently called upon to construe contracts and undertake to decide what the parties meant by the language of their agreement. When the language is clear and unambiguous, construction or interpretation is not required. But, when the language of the contract is ambiguous or obscure, the courts must seek to ascertain the intention of the parties. To find the parties' intention, the courts examine the language of the contract in the light of the nature, objects, and purpose of the contract and give the words their commonly accepted and reasonable meanings in the absence of evidence to a contrary intent. As Justice Oliver Wendell Holmes, Jr., so succinctly observed in *Towne v. Eisner*, 245 U.S. 418, 425 (1917): "A word is not a crystal, transparent and unchanged; it is the skin of a living thought and may vary greatly in color and content according to the circumstances and the time in which it was used."

So, in *Branagh and Sons v. Witcosky*, 242 Adv. Cal. App. 976 (1966), the contract stipulated the subcontractor would "fully indemnify and save harmless the contractor and owner against any and all loss, damage, liability . . . resulting from injury or harm to any person or property arising out of or in any way connected with the performance of the work under the subcontract, [excluding] only such injury or harm as may be caused solely and exclusively by the fault or negligence of contractor." During the performance of the contract, an employee of a second subcontractor was injured by the joint negligence of the first subcontractor and the contractor in violating an electric safety order.

Although the subcontractor objected to indemnifying the contractor on the grounds that it would encourage negligent conduct, the court held that the subcontractor was required to indemnify the contractor even though the injury resulted from the joint negligence of the contractor and subcontractor since the language of the contract was explicit and did not involve any ambiguity. The policy of seeking to uncover the parties' intent has resulted in a series of judicial priorities in weighing one circumstance against another.

Custom and Usage

Words have no inherent meaning but develop common meanings because they are used as tools of communication. While the courts will not make a contract where none exists, the court may look to the custom of the trade or industry to determine the true meaning of the contract. Technical words are given their technical meaning, and legal words are given their legal meaning. In addition, words may acquire special meanings based on use in a particular trade or industry (e.g., a "Christmas tree" is a complex of valves and pipes at the well head of a producing oil well). While the court could choose the dictionary's meaning of words, the parties' meanings usually take into account the setting and function of the transaction and other matters not included in a strictly dictionary definition. So, where a contractor agreed to dig ditches and lay pipe but the contract is silent on the obligation to backfill the trenches, the court considered the custom and usage in the pipeline industry to determine whether it was customary for the pipeline contractor or the owner to backfill the trenches.

Written Versus Printed

Another rule of construction distinguishes between written and printed portions of a contract. As a general rule, if the printed portion of the contract conflicts with the written (or typewritten) portions, the written or typewritten portions will control. This is based on the rationale that the parties may not have read the printed portions but that they have taken some care or special attention to prepare the written or typewritten portion.

Specific Provisions Control General Provisions

One observer of people in business noted that in negotiations many tend to talk in pleasant generalities, thinking they have arrived at an understanding yet failing to reach agreement on many, if any, unpleasant questions until forced to do so by their lawyers. So, rather than re-open negotiations, many business people deliberately accept an ambiguous provision hoping that the fuzzy understanding will not become an issue later.

Where the question involves a discrepancy between specific and general language, the rule of interpretation is that specific provisions or language control general provisions or language of the contract. By way of illustration, a housing project contract required "that the contractor perform all work necessary for completion of the project whether or not shown on the drawing" and then specified that the offsite electrical work be performed by a named electrical contractor. Although to properly finish the project it was necessary to install plumbing beyond the project limits, the contractor contended that such plumbing constituted "extra work." The court treating this as an exception to the general rule dealing with "offsite work" held that the contractor was not required to install the offsite plumbing since the parties had specifically excluded offsite electrical work.

Construing Vagueness Against Drafter

Under another principle of contract construction, ambiguous or uncertain provisions are construed most strongly against the party who drafted the instrument. The rationale of this doctrine is that if one party prepared the contract, an ambiguity in the contract should be construed against the drafter who had the opportunity to eliminate the ambiguity and failed to do so. This rule of contract construction is most frequently applied when a form contract is used and it can be reasoned that the party responsible for the contract form caused the uncertainty or ambiguity.

4.8   PAROL EVIDENCE RULE

Many contracts, at least complex agreements, are the result of numerous face-to-face meetings of various persons, telephone calls, letters, draft clauses, and memoranda before the final understanding emerges in a contract document. When a dispute occurs, the parties tend to review these earlier statements to interpret the meaning of the final document.

The parol evidence rule provides that oral or written extrinsic evidence of prior or contemporaneous agreements or negotiations is not admissible to contradict, vary, or modify an unambiguous written contract intended by the parties to be the final and complete expression of their agreement. The rationale of this rule is to decrease the possibility of awards based on perjured testimony and superseding agreements. However, the rule is severely restricted and does not operate to exclude the extrinsic evidence unless it is determined by the court that the parties intended the writing to be a final and complete expression or integration of their agreement.

Frequently the parties will include a contract clause stating that "[T]his writing is the complete expression of the agreement" or "[N]o other representations have been made." In the absence of fraud or mistake, such clauses are effective to establish the parties' intent to make a final and complete expression of their agreement. While the parol evidence rule excludes extrinsic evidence offered to contradict, vary or modify a fully integrated contract, such evidence may be offered and introduced to (1) show defects in the formation of the contract, (2) show alteration of the contract, or (3) show and explain an ambiguity. Perhaps most commonly, parol evidence will be introduced to translate the technical meaning of a word, explain a business custom, explain an ambiguity, or even show that because of fraud no valid contract exists.

The rule, which bars evidence of prior or contemporaneous agreements or negotiations, does not exclude evidence of subsequent agreements. It is a common problem in construction contracts that after the parties execute a contract the owner later makes changes in the work without written evidence of the changes. The parol evidence rule does not prohibit the contractor from introducing evidence to prove that the written contract has been modified. *Harrington v. McCarthy*, 420 P. 2d 790 (Idaho 1966).

NEWSOM v. UNITED STATES
676 F.2d 647 (U. S. Ct. of Claims 1982)

Smith, Judge: This case is an appeal by petitioner, [Newsom], of a decision of the Veterans Administration Board of Contract Appeals (board). The board found that certain parts of the contract for hospital improvements were patently ambiguous and that, having failed to consult with the contracting officer about the ambiguities, petitioner was barred from recovering for work done beyond that required under petitioner's interpretation of the contract. We affirm the decision of the board [denying petitioner's additional costs].

On August 28, 1978, the Veterans Administration (VA) issued an invitation for bids for building medi-prep and janitor rooms in the VA hospital at Knoxville, Iowa. Drawings and specifications for the work to be done were supplied to the prospective bidders.

Paragraphs 4, 5, and 6 of the specifications described, respectively, buildings 81, 82, and 85. Each paragraph had two parts: the first described the first floor of the building and referenced page 7 of the drawings; the second described the second floor of the building and referenced page 8 of the drawings. Conversely, the caption block on page 7 of the drawings indicated that it described work for all three buildings, 81, 82, and 85. However, page 8 of the drawings indicated only building 85. (It is not entirely clear whether a drawing of building 85 on page 8 of the drawings would have also described buildings 81 and 82, or whether separate drawings of buildings 81 and 82 were omitted from page 8. We do not believe that there is a difference of legal significance between these two possibilities.) Petitioner at no time inquired about this discrepancy.

As a consequence, petitioner included in his bid the costs of the second floor of building 85 only. He was the low bidder and the contract was awarded to him on October 13, 1978. It was not until [the following] March 29, that the parties realized that there was a discrepancy between what the VA had intended and petitioner had understood. Petitioner then did the work as intended by the VA at an additional cost of $14,000, and he appealed the decision of the contracting officer denying relief to the Veterans Board of Contract Appeals. The board held against petitioner on the ground that the error on page 8 of the drawings was a patent ambiguity which imposed upon the contractor a duty to inquire about it. Petitioner now appeals that finding to this court under the Contract Disputes Act.

The doctrine of patent ambiguity is an exception to the general rule . . . which requires that a contract be construed against the party who wrote it. If a patent ambiguity is found in a contract, the contractor has a duty to inquire of the contracting officer the true meaning of the contract before submitting a bid. This prevents contractors from taking advantage of the Government; it protects other bidders ensuring that all bidders bid on the same specifications; and it materially aids the administration of Government requiring that ambiguities be raised before the contract is bid on, thus avoiding costly litigation after the fact. It is therefore important that we give effect to the patent ambiguity doctrine in appropriate situations.

The existence of a patent ambiguity is a question of contractual interpretation which must be decided *de novo* by this court. This determination cannot be made upon the basis of a single general rule, however. Rather, it is a case-by-case judgment based upon an objective standard. In coming to our decision, we are bound neither by the legal conclusions of that board, nor by the subjective beliefs of the contractor, subcontractors, or resident engineer as to the obviousness of the ambiguity. The analytical framework for cases like the instant one was set out authoritatively in *Mountain Home Contractors v. United States*, 425 F.2d 1260, 1263 (1970). It mandated a two-step analysis. First, the court must ask whether the ambiguity was patent. This is not a simple yes-no proposition but involves placing the contractual language at a point along a spectrum: Is it so glaring as to raise a duty to inquire? Only if the court decides that the ambiguity was not patent does it reach the question whether a plaintiff's interpretation was reasonable. The existence of a patent ambiguity in itself raises the duty of inquiry, regardless of the reasonableness of the contractor's interpretation. It is crucial to bear in mind this analytical framework. The court may not consider the reasonableness of the contractor's interpretation, if at all, until it has determined that a patent ambiguity did not exist. [If the court finds ambiguity did not exist, then the reasonableness of the contractor's interpretation becomes crucial in deciding whether the normal rule to construe against the drafter applies.]

Examining the contract itself, we find that a patent ambiguity existed. Two parts of the contract said very different things: the specifications required construction on the second floors of buildings 81, 82, and 85, whereas drawings required construction on the second floor of only building 85. It is impossible from the words of the contract to determine what was really meant. The contractor speculated that it meant that part of the project had been dropped along the way. Looking at the same language, the Government can insist that it was clearly a drafting error. We do not consider which interpretation is correct; at this stage we determine only whether there was an ambiguity. What is significant about differing interpretations is that neither does away with the contract's ambiguity or internal contradiction. There is simply no way to decide what to do on the second floors of buildings 81 and 82 without recognizing that the contract also indicates otherwise.

The *Mountain Home* case, . . . involved a very similar ambiguity. The specifications ordered inclusion of kitchen fans in certain housing units, but the drawings appeared to indicate that kitchen fans were not to be installed.

Finally, we emphasize the negligible time and the ease of effort required to make inquiry of the contracting officer compared with the costs of erroneous interpretation, including protracted litigation. While the court by no means wishes to condone sloppy drafting by the Government, it must recognize the value and importance of a duty of inquiry in achieving fair and expeditious administration of Government contracts. Accordingly, upon consideration of the submissions, and after hearing oral argument, the decision of the Veterans Administration Board of Contract Appeals is Affirmed.

PARSONS v. BRISTOL DEVELOPMENT CO.
44 Cal. Rptr. 767, 401 P.2d 839 (1965)

Traynor, Chief Justice: In December 1960 defendant Bristol Development Company entered into a written contract with plaintiff engaging him as an architect to design an office building for a lot in Santa Ana and to assist in supervising construction. Plaintiff's services were to be performed in two phases. He completed phase one, drafting preliminary plans and specifications, on January 20, 1961, and Bristol paid him $600.

The dispute concerns Bristol's obligation to pay plaintiff under phase two of the contract. The contract provided that:

[a] condition precedent to any duty or obligation on the part of the OWNER (Bristol) to commence, continue or complete phase two or to pay ARCHITECT any fee therefor, shall be the obtaining of economically satisfactory financing arrangements which will enable OWNER, in its sole judgment to construct the project at a cost which in the absolute decision of the OWNER shall be economically feasible.

It further provided that when Bristol notified plaintiff to proceed with phase two it should pay him an estimated 25 percent of his fee, and that it would be obligated to pay the remaining 75 percent "only from construction loan funds."

Using plaintiff's preliminary plans and specifications, Bristol obtained from a contractor an estimate of $1,020,850 as the cost of construction, including the architect's fee of 6 percent. On the basis of this estimate, it received an offer from a savings and loan company for a construction loan upon condition that it show clear title to the Santa Ana lot [and] execute a first trust deed in favor of the loan company.

Shortly after obtaining this offer from the loan company, Bristol wrote plaintiff on March 14, 1961, to proceed under phase two of the contract. In accordance with the contract, Bristol paid plaintiff $12,000, an estimated 25 percent of his total fee. Thereafter, plaintiff began to draft final plans and specifications for the building.

Bristol, however, was compelled to abandon the project because it was unable to show clear title to the Santa Ana lot and thus meet the requirements for obtaining a construction loan. Bristol's title became subject to dispute May 23, 1961, when defendant James Freeman filed an action against Bristol claiming an adverse title. On August 15, 1961, Bristol notified plaintiff to stop work on the project.

Plaintiff brought an action against Bristol and Freeman to recover for services performed under the contract and to foreclose a mechanic's lien on the Santa Ana lot. The trial court, sitting without a jury found that Bristol's obligation to make further payment under the contract was conditioned upon the existence of construction loan funds. On the ground that this condition to plaintiff's right to further payment was not satisfied, the court entered judgment for defendants. Plaintiff appeals.

After providing for payment of an estimated 25 percent of plaintiff's fee upon written notice to proceed with phase two, paragraph 4 of the contract makes the following provisions for payment:

> 4. (b) Upon completion of final working plans, specifications and engineering, or authorized commencement of construction, whichever is later, a sum equal to SEVENTY-FIVE (75%) PERCENT of the fee for services in Phase 2 less all previous payments made on account of fee; provided, however, that this payment shall be made only from construction loan funds.

> (c) The balance of the fee shall be paid in equal monthly payments commencing with the first day of the month following payments as set forth in Paragraph 4(b) provided, however, that TEN (10%) PERCENT of the fee based upon the reasonable estimated cost of construction shall be withheld until thirty (30) days after the notice of Completion of the project has been filed.

> (d) If any work designed or specified by the ARCHITECT is abandoned or suspended in whole or in part, the ARCHITECT is to be paid forthwith to the extent that his services have been rendered under the preceding terms of this paragraph. Should such abandonment or suspension occur before the ARCHITECT has completed any particular phase of the work which entitles him to a partial payment as aforesaid, the ARCHITECT'S fee shall be prorated based upon the percentage of the work completed under that particular phase and shall be payable forthwith.

Invoking the provision that "payment shall be made only from construction loan funds," Bristol contends that since such funds were not obtained it is obligated to pay plaintiff no more than he has already received under the contract.

Plaintiff, on the other hand, contends that he performed 95 percent of his work on phase two and is entitled to that portion of his fee under subdivision (d) of paragraph 4 less the previous payment he received. He contends that subdivision (d) is a "savings clause" designed to secure partial payment if, for any reason, including the lack of funds, the project was abandoned or suspended. Plaintiff would limit the construction loan condition to subdivision (b), for it provides "that *this payment* shall be made only from construction loan funds" [emphasis added], whereas the other subdivisions are not expressly so conditioned.

The construction loan condition, however, cannot reasonably be limited to subdivision (b), for subdivision (c) and (d) both refer to the terms of subdivision (b) and must therefore be interpreted with reference to those terms. Thus, the "balance of the fee" payable "in equal monthly payments" and subdivision (c) necessarily refers to the preceding subdivisions of paragraph 4. In the absence of evidence to the contrary, subdivision (d), upon which plaintiff relies, must likewise be interpreted to incorporate the construction loan condition . . . for it makes explicit reference to payment under preceding subdivisions by language such as "under the preceding terms" and "partial payment as aforesaid." Subdivision (d) merely provides for accelerated payment upon the happening of a contingency. It contemplates, however, that construction shall have begun, for it provided payment upon the abandonment or suspension in whole or in part of "any work designed or specified by the Architect." Implicit in the scheme is the purpose to provide, after initial payments, for a series of payments from construction loan funds, with accelerated payment from such funds in the event that construction was abandoned or suspended. Although plaintiff was guaranteed an estimated 25 percent of his fee if the project was frustrated before construction, further payment was contemplated only upon the commencement of construction.

Accordingly, the trial court properly determined that payments beyond an estimated 25 percent of plaintiff's fee for phase two were to be made only from construction loan funds. [For defendant, Bristol Development Co.]

DISCUSSION QUESTIONS

1. Illustrate fact situations which demonstrate the difference between (1) donee-beneficiary, (2) creditor-beneficiary, and (3) incidental-beneficiary. What are the rights of each category of beneficiary against the other parties?

2. Discuss the murky area between unilateral mistake and the failure of a contracting party to seek clarification of the contract terms.

3. What is the legal status of an oral contract found to be within the Statute of Frauds? Describe factors and evidence that would take the oral contract out of the Statute of Frauds.

4. Distinguish between assignment of a contract and a third-party beneficiary.

5. Buyer orders by telephone a quantity of diesel at an agreed per barrel price from Seller. Buyer is thinking Los Angeles delivery and Seller is thinking San Diego delivery as they talk on the telephone, and their testimony is in conflict as to the destination discussion. Promptly after the conversation, Seller sends a telex to Buyer confirming the elements of the transaction stating San Diego delivery and asking for prompt advice if the terms are not as agreed. Upon receiving the telex, the Buyer checks price, quantity, and quality but does not notice the incorrect destination. In turn, Buyer sends a telefax confirmation, which states Los Angeles as delivery point but that telefax is not delivered to Seller. On the date to nominate delivery, the parties discover this error. What results and why?

5

BREACH OF CONTRACT
Excuse and Remedies

5.1 BREACH OF CONTRACT

In determining whether a contract has been breached, the court examines the contract (whether written or oral), inquires into any antecedent negotiations that were not superseded by the actual contract, and looks for any terms implied by law.

Conditions Precedent and Subsequent

Frequently, performance by a party is contingent upon the occurrence of an event. This event, legally called a "condition," is classified as either a condition precedent or a condition subsequent. Where no duty to perform comes into existence until a specified external event occurs (e.g., the obligation to purchase is subject to the buyer's obtaining satisfactory financing), that event is a condition precedent. Failure of the buyer to obtain satisfactory financing will excuse the buyer's performance to purchase. A condition subsequent also excuses performance but involves the happening of an external event after the duty to perform has been established (e.g., where the insurance obligation to pay for a loss will be excused when notice of loss is not given under an insurance policy requiring such notice within 15 days after the occurrence of the loss).

Performance and Breach

While the truism that full performance of the agreement discharges the contractual duties, its corollary whether less than full performance is or may be a breach is the source of considerable litigation.

The courts recognize three stages of performance: (1) complete and satisfactory performance, (2) substantial performance, and (3) material breach. Complete and satisfactory performance represents full performance or to a degree that nothing else could be reasonably expected. Substantial performance implies an honest effort to perform, but a performance of less than expected; and these shortcomings in performance are dealt with in the parties' remedies. If there is a major defect in performance, there has been a material breach for which the breaching party will be required to pay damages to make whole the nonbreaching party. The measure of damages is discussed later in this chapter.

The issue of substantial performance is common in construction contracts for a number of reasons. First, construction is a complex undertaking, and it is not uncommon for minor deviations to surface at the completion of the project. Second, work performed by the subcontractor's employees may escape the scrutiny of the prime contractor's inspectors or supervisory employees. Third, not only is much of the work performed by subcontractors but the contract documents are complex and frequently generate questions of interpretation that are difficult to answer while the work is in progress. Finally, the performance of the contractor, after it has been incorporated into the owner's land, cannot be retracted. For example, in *Jacobs & Youngs Inc. v. Kent*, 129 N.E. 889 (N.Y. 1921), the court found that there had been substantial performance, although one brand of pipe (which the court found to be of equal value) had been substituted for the brand specified in the contract.

Time of Performance

A failure to perform in a specified time usually will not be considered a material breach, especially if there is some justification for the delay. In *Kole v. Parker Yale Development Co.*, (Ct. App. Colo. 1975), where plaintiff owner sought rescission because condominium units were not completed within the agreed contract time and the agreement contained no stipulation concerning that time was of the essence, the court found time was not of the essence and denied rescission, saying, "damages provided an adequate remedy." Nevertheless, because of the interrelated scheduling of work on various construction contracts, a delay may result in costly rescheduling that may have to be dealt with in measuring damages. So, where time is not of the essence, then performance completed within a reasonable time must be accepted, although the nonbreaching party may recover damages for the late performance. On the other hand, if time of performance is essential to proper performance of the contract, the failure to perform within the expected time is a material breach of the contract. In contracts in which performance at a given time is significant, it is customary to include a clause such as "time is of the essence in this contract."

5.2 ANTICIPATORY BREACH

A breach usually occurs when a party fails to perform as agreed or within a reasonable time set forth in the contract. However, prior to the time for agreed performance, a party may announce an intention not to perform. Such action is referred to as "anticipatory breach" or "repudiation," and the nonbreaching party may treat such action as an immediate complete breach and proceed to bring suit for damages or other appropriate relief.

When an anticipatory breach occurs, the party entitled to performance has the choice of several courses of action. First, an action for breach may be brought at once to recover damages for the breach. In this case, as a condition of recovery, there must be proof that the party bringing suit was ready and willing to perform. Second, if the party electing to sue has performed in whole or in part, an action may be brought seeking restitution. This remedy is essentially a rescission of the contract.

In a construction project situation, a contractor has the option either to affirm the contract and recover all damages sustained because of the breach, including the profit the contractor could have made if permitted to complete the contract, or, in the alternate, to rescind the contract and recover the reasonable value of the benefits conferred on the breaching party for the work and materials already furnished. In the event of the contractor's breach, the owner may eject the contractor from the job and hire another contractor to finish the work and hold the original contractor liable for damages measured by the difference between the contract price and the actual cost of completion.

Similarly under Section 2-712 of the Uniform Commercial Code (UCC), a buyer of goods may "cover" in good faith by purchasing, without unreasonable delay, goods in substitute of those to be furnished by the breaching party and recover damages for any higher cost of the substitute goods purchased. Failure of the party to cover will not bar damages for the breach; however, under the duty to mitigate, the recoverable damages will be limited to those that could have been obviated by proper timely cover.

5.3 IMPOSSIBILITY OF PERFORMANCE

In 1467, an English court in the case of *Pardine v. Jane*, 82 Eng. reprint 897, announced the rule of law that where a party by his or her own contract creates a duty, that party is bound to performance notwithstanding any accident or inevitable necessity since the party is free to contract against such risks. Over the years, this strict doctrine has been somewhat modified by judicial decisions so that actual impossibility of performance is a valid excuse for breach of contract and releases a party from the duty to perform. However, true impossibility is rare, and what is usually asserted as a defense for nonperformance is, in fact, that performance cannot be accomplished without excessive and unreasonable cost.

The question has been posed as to whether the unanticipated circumstance has made performance of the promise vitally different from what should reasonably have been within the contemplation of both parties when they entered into the contract. If so, the risk fairly should not be thrown upon the promisor. Professor Corbin, having noted that many supervening events affect the relative values of the agreed performances, wrote:

> Up to a point, such events . . . do not cause contractors to ask or to expect relief. But there comes a point at which the request is regarded as reasonable and the expectations will not be disappointed. Where is that point? Unfortunately, it is a vanishing point; the line that it draws is an invisible line. In any specific case, its location depends upon customary business practice and commonly prevailing opinion. 6 Corbin, *Contracts*, Section 1354, at p. 459 (1962).

To determine whether an event that makes performance impossible discharges the duties of the parties, poses the issue of what risks, given the knowledge and experience of the parties, were reasonably contemplated by the contract. If the risk was not reasonably foreseeable at the time of formation of the contract and was not allocated by the contract, occurrence of the event will discharge each party's performance. Occurrences, such as (1) strikes in the party's own performance,

(2) unavailability of materials, and (3) government regulations that make performance more difficult are normally foreseeable.

In *Caron v. Andrew*, 284 P.2d 544 (Cal. 1955), a contractor agreed to level 755 acres of land by a given date. During performance of the work, a levee broke, and the land was flooded. It was common knowledge that the land was subject to inundation. The occurrence of floods at the particular season, although unusual, was not unknown, and the contract did not provide for an extension. The court held the contractor liable saying that the contractor, having agreed to complete the work by the designated date, assumed the risk of delay or expense because of flooding.

Closely related to the excuse of impossibility is the unforeseen occurrence of an extraordinary event (frequently referred to as *force majeure* or "acts of God"), thus changing the underlying assumptions so that it is unjust to hold the party to performance. In the U.S. legal system it is customary to include such a clause with a laundry list of events (e.g., floods, tornados, lightning, winds, hail, government regulation, war, insurrection, strikes) that will excuse or at least postpone the required performance.

5.4 COMMERCIAL IMPRACTICABILITY: UCC

In addition to a mellowing of the judicial policy toward impossibility of performance, the Uniform Commercial Code makes provision for several situations in which the promised performance in connection with the sale of goods becomes impracticable. It is important to note there is a difference between "impossibility" and "impracticability," which may be compared with frustration. Dealing with the matter of commercial impracticability in the sale of goods, Section 2-615(a) of the UCC rejected the strict requirements of the law of impossibility in certain respects. The UCC provided that failure to perform in the delivery of goods under a contract may be excused where performance ". . . has been made impracticable by the occurrence of a contingency, the occurrence of which was a basic assumption on which the contract was made." In that situation, if the seller is able to furnish only a portion of the goods, the seller is required ". . . to allocate production and deliveries among customers, in any manner which is fair and reasonable, and the seller may elect to include regular customers not then under contract as well as seller's own requirements for further manufacture." Section 2-615(c) of the UCC also requires that the seller ". . . notify the buyer seasonably that there will be delay or non-delivery and, when allocation is required [must notify the buyer] of the estimated quota which will be made available to the buyer."

5.5 REMEDIES

In General

When there is no excuse for nonperformance (such as fraud, failure of conditions, or impossibility), a party has an absolute duty to perform. Failure to perform, in addition to excusing the other party's duty to performance, will entitle the innocent party to some affirmative relief. Justice Oliver Wendell Holmes, Jr., speaking of damages, declared "the duty to keep a contract at Common Law means a prediction that you must pay damages if you do not keep it." However, in such cases, the nonbreaching party may have several alternate remedies available such as (1) damages, (2) specific performance, (3) rescission and restitution, (4) reformation, and (5) quasi-contract damages for unjust enrichment. Some of these remedies may be more favorable to the nonbreaching party than other possible remedies. Likewise, in some cases, the breaching party may find nonperformance and the payment of damages less onerous (in fact, an incentive at times) rather than the cost of continuing performance.

Damages

The award of money damages is the normal method of compensating a nonbreaching party for the injuries sustained by the other party's failure to perform. The purpose of damages is to place the nonbreaching party in the position that the party would have occupied had the contract been fully performed. The law of damages protects three basic interests: (1) restitution interest for benefits conferred, (2) reliance interest for out-of-pocket expenses incurred in preparing to perform, and (3) the expectation interest that the party would receive the benefit of the contract bargain. The objective of any of these interests is the underlying policy justifying recovery.

Nominal and actual damages. The breach of a contract entitles the injured party to recover general damages which may be either nominal or actual. Nominal damages, normally a trifling sum, are awarded in recognition of a technical breach and the plaintiff's right to recover. This limitation to nominal damages may be due to the failure to prove actual damages or situations where the actual damages are too speculative to form a basis for recovery. On the other hand, actual damages ("compensatory damages") are intended to place the nonbreaching party in the position that the injured party would have enjoyed had the contract been performed.

In partially completed contracts, the court frequently uses a formula to determine the amount of actual compensatory damages by predicting what would have happened and awarding the injured party the contract price less (1) any progress payments and (2) the cost of completion, which is the expense saved by the breach.

This measurement of the "expectation interest" requires a court to determine what the parties' positions would have been had there been no breach. Ordinarily, if a contractor in building a porch for an owner performs the work defectively, the owner may recover the cost of needed repairs to fully remedy the defects. However, this analysis presents special problems in construction cases where the cost to complete or repair far exceeds any diminution of the reasonable value of complete performance. Such a situation was presented in *Eastlake Construction Co. v. Hess*, 686 P.2d 465 (Wash. 1984), where the court held that the cost of corrective work is the proper measure of damages unless this cost is disproportionate. If the cost is disproportionate, then the measure of damages is the difference between the market value of the project with and without the defect. In that case, it may be better to use a diminution of value theory. In another case, *Triad Constructors Inc. v. Morris*, 214 S.E.2d 209 (N.C. 1975), where a contractor constructed a concrete flow too low to the ground on a sloping lot with the result that water would not drain properly and actually ran into the building, the cost of repair far exceeded the diminution in market value caused by the water problem. There the court held that damages should be measured by the diminution in value to the owner as reflected in the decrease in rental value of the property owing to the water problem.

The plaintiff, as a part of its proof, must present evidence as to the amount of the alleged damages with a reasonable degree of certainty since damages will not be awarded that are too "remote and speculative." While subscribing to this rule, the courts have seemingly viewed fact situations differently. A Florida court, in a case involving delay in construction completion, refused lost profits from rents. On the other hand, a California court in *Cooper v. Jevne*, 56 Cal. App. 3rd 860, 128 Cal. Rptr. 724 (1976), permitted recovery of "loss of use and income" owing to the alleged malpractice by an architect.

Special or consequential damages. Special or consequential damages are those damages that do not flow directly from the breach of contract but are due to special circumstances of the contract. In assessing actual compensatory damages, most jurisdictions follow the rule of the early English case of *Hadley v. Baxendale*, which established that only those damages are recoverable that were the necessary, natural, and probable result of the breach or which were reasonably within the contemplation of both parties at the time of entering the contract. In that case plaintiff Hadley contracted with common carrier Baxendale to have Baxendale deliver Hadley's broken mill shaft to a foundry to be repaired. Because Baxendale delivered the shaft several days late, Hadley had to close down the mill and sued Baxendale for loss of profits resulting from the mill's closing. While Baxendale would have been liable for damage to the shaft while in transit, the court held the closing down of Hadley's mill was not a reasonably foreseeable result of Baxendale's breach.

The courts generally limit a contracting party's responsibility to a standard of foreseeability. This rule is based on the theory that the damages, in the absence of notice, cannot be considered fairly to have been within the contemplation of the parties as part of the consequences flowing from a breach when the contract was made. The case of *R.E.T. v. Frank Paxton Co.*, 329 N.W.2d 416 (Iowa 1983), involved an action brought by an owner against the contractor, based on the failure to comply with the insulation requirements of the contract in an apartment complex. Tenants' complaint concerning cold apartments and high heating bills required rent concessions by the owner during the 3 year period in which the owner was completing the necessary corrective construction. As a result, the complex fell into serious financial difficulties and was ultimately sold at a loss. The trial court awarded damages of $105,000 for repair costs, $237,000 for lost rents, and $650,000 based on the loss in the sale after finding breaches of express and implied warranties as well as negligent installation of the insulation. The Iowa supreme court in affirming the trial court's award of damages commented that the trial court had erroneously concluded that the damages are the same for breach of contract and for negligence. The court held that the cost of correction and an amount for the diminished value would be necessary to compensate the plaintiff since the parties could have contemplated the cost of repairs and lost rentals. Finally, although the court reasoned that while contracting parties would not normally contemplate diminution of value and the resulting distress sale, such damages would be recoverable in a claim based on tort and thus there was no reversible error. The distinction in the theory of damages resulting from a tort injury is discussed in chapter 7.

Liquidated damages. In anticipation that the precise amount of damages may be difficult to determine, the parties may contractually agree on specific sums to be paid in the event of breach. The term "liquidated damages" refers to a specific amount designated by the parties at the time of contracting, which is recoverable by the injured party in event of breach of the contract. Since liquidated damages are in the nature of a forfeiture, which is disfavored by the courts, they cannot be recovered unless (1) the damages anticipated are uncertain in amount or difficult to prove, (2) the parties tailored the liquidated damages to the circumstances of the contract in advance, and (3) the agreed amount is reasonable and not so large as to be considered a penalty.

The court must be able to find that at the time of the initial agreement it appeared to the parties that the actual losses that would result from the breach would be extremely difficult or impossible to estimate. Illustrative of this requirement is a contract to finance the purchase of $9,000,000 of jet aircraft that provided for $250,000 liquidated damages to the lender in the event that the aircraft purchase failed through no fault of the lender. The court, in *Walter E. Heller & Co. v. American Flyers Airline Corp.*, 459 F.2d 896 (2d Cir., 1972), upheld the clause since $9,000,000 was involved in the transaction and there was no easy method of calculating the loss to the lender from the inability to finance the planes.

In dealing with the requirement that the amount of liquidated damages must represent a reasonable effort to predict actual damages, a liquidated damage clause providing for $50 per day of delay beyond completion date set for construction of a sewage treatment facility was held reasonable in *Louis Lyser General Contractor Inc. v. City of Las Vegas*, 489 P.2d 646 (N.M. 1971). The clause was enforceable in the light of the $109,000 total cost of the project and the seriousness of depriving several thousand people of sewage treatment facilities.

Punitive damages. Normally, punitive damages are not awarded in addition to compensatory damages for breach of contract. However, where the breach is malicious or intentionally harmful, bordering on a "tort action," the courts have occasionally allowed punitive damages.

A Wisconsin case, *Jeffers v. Nysse*, 297 N.W. 2d 495 (1980), awarded punitive damages against a developer-builder who misrepresented the insulation and heating costs of a house. Although the jury failed to find that the builder acted with malice, it found the conduct had been wanton and reckless. The court concluded,

> mere compensation for their [the purchaser's] costs in putting the house in the condition it was represented to be originally, equated to putting the cookies back in the jar when caught. This was not enough. If that result were reached, seller could make any misrepresentations necessary to make a sale. If it was not discovered, or was discovered and not pursued, the seller would make a windfall gain. If the fraud were discovered and successfully proven, the seller would only be liable to make good on his representations. He would suffer no punishment nor would he be deterred from similar conduct in the future.

Mitigation of damages. The injured party owes a duty to make a reasonable effort to mitigate damages and to minimize the damages that will be sustained by the injured party. This rule limiting damages relates to the requirement that the breaching party is responsible only for those losses that its breach has caused. The injured party must take reasonable steps (described in Section 350 of the RESTATEMENT of Contracts as those which do not impose ". . . undue risk, burden or humiliation . . . .") to reduce the actual loss to a minimum. For example, an employee who has been wrongfully discharged cannot sit idly by and expect to draw pay. A duty is imposed to seek other work of substantially similar character in the same community; but, the employee is not required to accept employment of a different or an inferior kind. See *Parker v. Twentieth Century Fox Film Corporation*, 89 Cal. Rptr. 737 (Cal. 1970). In the case of breach by the owner, a contractor was obligated to stop purchasing supplies and use or sell materials already purchased. Although it is unusual for the court to challenge the cost of completion after breach by a contractor, in *First National Bank of Akron v. Cann*, 503 F. Supp. 419 (1980) affirmed 669 F.2d 415 (6th Cir. 1982), the court reduced the award on the basis that overtime was not necessary and the claimant's refusal to allow the designer-builder to provide free engineering services was not justified.

Rescission and Restitution

Rescission of the contract is an appropriate remedy where one of the parties lacked capacity to contract or there was a failure of consent owing to fraud or undue influence. The effect of rescission is to put the contract to an end for all purposes and so far as possible to restore the parties to their original positions. Accordingly, a party seeking rescission of the contract cannot later maintain an action for breach of the contract. The party discovering facts that warrant rescission of a contract has the duty to act promptly, and, if rescission is elected, to notify the other party within a reasonable time so that rescission may be accomplished at a time permitting the parties to be restored, as nearly as possible, to their original positions. A party, although entitled to rescission, may also elect to affirm the contract and seek to recover damages, but, once that choice to affirm is made, the right to rescind is lost.

In granting the equitable remedy of rescission, the court, after terminating the contract and restoring the parties to their original positions, may award money damages in restitution. For example, serious breaches by the owner such as failing to make progress payments, making excessive changes, or refusing to perform acts that allow the contractor to proceed in the most expeditious way may be grounds for the contractor to terminate the contract and seek rescission and restitution. Because rescission looks backward to restoring the parties to their position before the contract, this may be impossible in construction situations where improvements have been attached to the breaching owner's land. In that case, the contractor is entitled to receive the "value," in terms of unjust enrichment under quasi-contract that its partial performance benefited the owner. Ordinarily, the contractor will introduce evidence of contractor's actual costs incurred in performance (as evidence of the benefit to the owner) and, while the contractor cannot recover profit on unperformed work, profit on work performed has been allowed. *Leo Spear Construction Co. v. Fidelity & Casualty Co. of New York*, 446 F.2d 439 (2d Cir. 1971).

The underlying theory of a quasi-contract is based on the equitable principle that it would be unfair for a person having received a benefit from another to keep or enjoy the benefit without paying for it. An agreement is implied in law based on three essential elements: (1) benefit conferred, (2) appreciation or knowledge of the benefit, and (3) retention of the benefit under circumstances that make it unfair to retain the benefit without paying. In *Gebhardt Brothers, Inc. v. Brimmel*, 143 N.W.2d 479 (Wisc. 1966), a subcontractor under an agreement with the prime contractor supplied fill dirt for the construction site on the developer's land for which the developer paid the prime contractor but the prime contractor failed to pay the subcontractor. The subcontractor's suit in quasi-contract against the developer failed when the court concluded that the last of the three elements was missing since the developer had paid for the benefit.

In dealing with situations of quasi-contracts, a nonbreaching contractor may elect to sue under quasi-contract for the reasonable value of the work performed. In this manner, the recovery may exceed that contract price, although there is a line of cases limiting recovery to the contract price. A breaching contractor may recover for the value of services rendered only if the contractor has substantially performed or if the owner accepts defective performance.

Specific Performance

Courts may order specific performance of a contract when a damage remedy would be inadequate. The equitable remedy of specific performance is an injunction (an order of the court for a party to do or not to do a particular act) that is available to prevent or protect against hardship where the legal remedy of dollar damages is inadequate. Dollar damages are considered inadequate and specific performance is the proper remedy when the subject matter of the contract is unique, such as contracts for the sale of land, antiques, or art objects and where it would be difficult to place a value on the item or to acquire a satisfactory substitute with the damages awarded. Contracts for personal services will not be specifically enforced or where the decree would require the prolonged supervision by the court of conduct such as the construction of a building, the remedy of specific enforcement is seldom granted.

MALO v. GILMAN
379 N.E.2d 554, (Ind. 1978)

Gilman, who contemplated construction of an office building, engaged Malo, an architect, to complete the necessary plans and specifications. The construction bids received totaled $105,000, 50% more than the preliminary estimated cost of $70,000, which appeared in the contract. Gilman was unable to secure financing and the building was never built. Malo brought an action to recover his fee as architect. Gilman counterclaimed for the $500 he had paid to Malo. The trial court found for defendant Gilman and granted his counterclaim. We affirm the judgment.

Malo agreed to provide architectural services to Gilman in the design and construction of an office building. Between 1967, and November, 1968, Malo expended considerable time and effort on the project. A verbal agreement was reached on July 28, 1967. Gilman was assured that costs of construction could be kept below $20.00 per square foot. After talking to prospective tenants, Gilman decided he required 3,500 square feet in the building. A standard American Institute of Architects (AIA) form contract was signed on May 14, 1968. Among the terms were the following:

It is recognized that this written contract ratifies the similar verbal contract entered into July 28, 1967.

The preliminary estimated cost of this project is Seventy Thousand Dollars, ($70,000).

Statements of Probable Construction Cost and Detailed Cost Estimates prepared by the Architect represent his best judgment as a design professional familiar with the construction industry. It is recognized, however, that neither the Architect nor the Owner has any control of labor, materials or equipment, or over the contractors' methods of determining bid prices, or over competitive bidding or market conditions. Accordingly, the architect cannot and does not guarantee that bids will not vary from any Statement of Probable Construction Cost or other cost estimate prepared by him.

Malo completed the plans and specifications for the project in September, 1968. In October, 1968, bids were solicited. The lowest total of bids received was approximately $128,000, which was negotiated down to $105,000. The bids were never accepted. Gilman indicated that the bids were unacceptable to him and that he was unable to secure financing. In December, Gilman sold the land on which the building was to have been erected.

Malo demanded payment of his fee for architectural services in the sum of $9,132.60. Gilman refused to pay. Malo brought an action to collect his fee. Gilman counterclaimed for the $500 he previously had paid Malo.

We hold that the judgment of the trial court [which denied Malo's claim but granted Gilman's counterclaim to recover money already paid to Malo] can be affirmed on either of two alternate theories: (1) that parol evidence was properly admitted to show a maximum cost limitation of approximately $70,000, which was exceeded unreasonably by Malo's plans for construction; or (2) that the estimated cost figure appearing in the contract placed a reasonable limit on the actual cost of the project, which limit was exceeded unreasonably. In either event, architect Malo breached the contract and is not entitled to compensation under the contract.

I. Parol evidence to show a maximum cost limitation. On appeal, Malo argues that no fixed price agreement appeared in the "fully integrated contract" for architectural services. The only figure appearing in the contract, $70,000, was merely a preliminary estimated cost figure which was not binding on the architect. Further, even if the trial court properly allowed evidence of a $20 per square foot cost limitation, Malo claims his final design plans contained 5,400 square feet, 50% more space than originally projected. In that case, the bids totaling $105,000 were in the right price range for a building costing $20 per square foot.

Gilman contends that evidence showing the existence of a $20 per square foot cost limitation (or $70,000 to $78,000 total for the project) was properly admitted, since the contract failed to contain a maximum cost limitation. Further, no significant changes in the project occurred to increase its size or cost.

Normally parol evidence may not be considered if it contradicts or supersedes matters intended to be covered by the written agreement. However, parol evidence may be admitted to supply an omission in the terms of the contract. Many contracts for architectural services, as here, fail to include specific requirements such as the size, style, and character of the building, the number of rooms, the quality of the materials to be used, and, finally, the maximum cost. [The same AIA form contract] provides:

> The Architect shall consult with the Owner to ascertain the requirements of the Project and shall confirm such requirements to the Owner.

Thus, depending on the specific needs of the owner, these requirements may be integral parts of the contract for architectural services. A contract that fails to set out the details agreed upon, then, is not a complete and integrated statement of the agreement. Parol evidence may be considered to determine the agreement with respect to these matters.

Ordinarily, the maximum cost of a project is agreed upon prior to commencement of design. The owner who plans to construct a building has in mind a figure for the maximum cost of construction, particularly where, as here, he must secure outside financing. The architect must design the project, keeping in mind this maximum cost limitation. Evidence of the maximum cost limitation should be admissible where the contract fails to show that figure.

We agree that parol evidence of a maximum cost limitation may be introduced where the contract fails to contain such a limitation. The question of fact, whether architect Malo agreed to design a building, the cost of which would not exceed $20 per square foot (or $78,000 for 3,900 square feet), was resolved by trial court in favor of Gilman. On examining the record, we cannot say that the finding of fact was incorrect as a matter of law.

Gilman testified that, from the beginning, he received repeated assurances that Malo would have "no problem" designing a building costing less than $20 per square foot. The amount of usable space Gilman required was approximately 3,500 square feet. By multiplying the figures, Gilman and Malo arrived at $70,000 as a "firm figure" for maximum cost. Gilman sought to reduce costs, accepting the use of a cost-saving "Uni-Roof" design, and suggesting a cost-cutting relocation of the basement.

Malo testified that he was aware that Gilman was interested in "getting the best price on the market." Yet he claimed there was no ceiling on the cost of the project, that "it could be as much as a million dollars." Finally, one of plaintiff's witnesses, who was present at a meeting where the bids were tabulated, testified that no specific "cost talk" was discussed, that he did not recall the figure of $70,000, but that $80,000 kept "ringing a bell."

The evidence fails to show that Gilman would have paid *any* sum of money to construct the proposed building. On the contrary, the evidence clearly shows that he wished to construct the building as cheaply as possible. The trial court correctly resolved the factual question of a maximum cost limitation in favor of Gilman. Under the terms of the agreement, then, Malo lost his right to recover compensation when he designed a building impossible of construction within the maximum cost limitation.

II. Estimated cost figure exceeded unreasonably. The contract for architectural services contained an estimated cost figure of $70,000. After negotiation, the lowest construction bids totaled $105,000, a figure 50% higher. Appellee Gilman argues that so great a discrepancy should bar Malo from receiving his fee for architectural services. [An earlier case supports] such a theory where the building would cost at least $57,000 to build, or $12,000 (almost 30%) in excess of the cost limitation on the project. The court concluded,

> that plaintiff has not substantially complied with the terms of his written contract and, therefore, is not entitled to recover in this action....

[In a recent case involving construction bids which exceeded the estimate by 55%, the court] interpreted the section of the contract in which the architect explicitly refused to guarantee the cost estimate, declaring that the figure represented a "reasonable approximation of the cost of the project." This did not mean that an architect would never be bound by his estimate. Such a situation would be "contrary to public policy because it would mean that no matter how large the bid for doing the work, defendants would be obligated to pay an architectural fee based on that amount . . . ." The court held:

> that an architect or engineer may breach his contract for architectural services by underestimating the construction costs of a proposed structure. The rule to be applied is that the cost of construction must reasonably approach that stated in the estimate unless the owner orders changes which increase the cost of construction . . . .

III. Conclusion. Under either theory, the trial court could have found that Malo breached his contract for architectural services and denied Malo compensation under the contract, and found that Gilman was entitled to recover the $500 he paid Malo under the contract. The judgment of the trial court is affirmed.

BARRON G. COLLIER, INC. v. DEUTSER FURNITURE CO.
256 S.W.300 (Tex. 1923)

Action by Barron G. Collier. Inc. [plaintiff], against B. Deutser Furniture Company [defendant]. The plaintiff, which company through various contracts controlled all advertising space in streetcars in the city of Houston, Texas, entered into two contracts dated June 1, under which the defendant agreed to rent a specified number of advertising spaces for a period of five years from the date of the contracts. On August 2 [of the following year], defendant requested plaintiff to cancel and rescind the contracts. After the last-mentioned date, defendant refused to use the advertising space called for in the contract or make any further payments. The plaintiff brought this action for a breach of the contracts and sought to recover as damages the contract price of the advertising space for the remainder of the contract period.

Hightower, Chief Justice: The undisputed evidence in this case shows that on July 2, defendant duly notified plaintiff by letter that it would not use any of the space contracted for in any of the streetcars after August 1, and plaintiff was directed by the letter to re-rent such space in said cars or make such disposition of same as it might wish. Having been thus notified by defendant that it would no longer comply with its written contracts to use the space in any of the cars after August 1, it was plaintiff's duty to use reasonable efforts and diligence to ward off or minimize such threatened damages as might flow from defendant's breach of contract, and if plaintiff failed to use reasonable efforts and diligence to minimize such threatened damages, as was found by the trial court it did, it was not entitled to recover for any such damages as it might have prevented by the use of such efforts and diligence. It seems to be the general rule . . . that one who is threatened with damages in consequences of a tort or breach of contract, with some few special exceptions, must use reasonable diligence and efforts to ward off or minimize such threatened damages.

We think, from the testimony of plaintiff's manager, Smiley, as reflected by this record, that plaintiff could have re-entered or relet the Deutser space in the local cars for as much or more than it was to receive under the Deutser contracts during the entire period of time between the breach of the contract and the trial of the case. [For Appellant Defendant]

BETHLEHEM STEEL CORPORATION V. CHICAGO
350 F.2d 649 (7th Cir. 1965)

Grant, District Judge. Plaintiff [Bethlehem] brought this action to recover an item of $52,000 together with certain items of interest, etc., withheld by the Defendant [city of Chicago], as liquidated damages for delay in furnishing, erecting, and painting of the structural steel for a portion of the south route superhighway, in the city of Chicago. [T]he District Court concluded that Plaintiff's claims on the items in controversy should be denied and entered judgment accordingly. We agree and affirm.

The work which Bethlehem undertook was the erection in Chicago of structural steel for a 22-span steel stringer elevated highway structure, approximately 1,815 feet long, to carry the South Route Superhighway from South Canal Street to the south Branch of the Chicago River. Bethlehem's work was preceded and followed by the work of other contractors on the same section. The Proposal and Acceptance in the instructions to bidders required the bidders to ". . . complete . . . within the specified time the work required . . . ." Time was expressly stated to be the essence of the contract and specified provisions were made for delivery of the steel within 105 days thereafter, or a total of 195 days after commencement of work, which was to be not later than 15 days from notification. The work had to be completed irrespective of weather conditions.

The all important provision specifying $1,000 a day "liquidated damages" for delay is as follows:

The work under this contract covers a very important section of the South Route Superhighway and any delay in the completion of this work will materially delay the completion of and opening of the South Route Superhighway thereby causing great inconvenience to the public, added cost of engineering and supervision, maintenance of detours, and other tangible and intangible losses. Therefore, if any work shall remain uncompleted after the time specified in the Contract Documents for the completion of the work or after any authorized extension of such stipulated time, the contractor shall pay to the City the sum listed in the following schedule for each and every day that such work remains uncompleted, and such moneys shall be paid as liquidated damages, not a penalty, to partially cover losses and expenses to the City.

[Schedule of Liquidated Damages]

"Amount of Liquidated Damages per Day—$1,000

Bethlehem's work on this project followed the construction of the foundation and piers of the superhighway by another contractor. Bethlehem, in turn, was followed by still another contractor who constructed the deck and the roadway.

Following successive requests for extensions of its own agreed completion date, Bethlehem was granted a total of 63 days' additional time within which to perform its contract. Actual completion by Bethlehem, however, was 52 days after the extended date . . . . [Bethlehem appeals an assessment by the City of] $1,000 per day, or a total of $52,000 as liquidated damages.

Bethlehem contends it is entitled to the $52,000 on the ground that the City actually sustained no damages. Bethlehem contends that the above-quoted provision for liquidated damages is, in fact, an invalid penalty provision. It points out that notwithstanding the fact that it admittedly was responsible for 52 days of unexcused delay in the completion of its contract, the superhighway was actually opened to the public on the date scheduled.

In other words, Bethlehem now seeks to re-write the contract and to relieve itself from the stipulated delivery dates for the purposes of liquidated damages, and to substitute therefor the City's target date for the scheduled opening of the superhighway. This the plaintiff cannot do.

[The Supreme Court in *Wise v. United States*] stated: "[T]he courts will endeavor, by a construction of the agreement which the parties have made, to ascertain what their intention was when they inserted such a stipulation for payment, of a designated sum or upon a designated basis, for a breach of a covenant of their contract . . . . When that intention is clearly ascertainable from the writing, effect will be given to the provision, as freely as to any other, where damages are uncertain in nature or amount or are difficult of ascertainment or where the amount stipulated for is not so extravagant, or disproportionate to the amount of property loss, as to show that compensation was not the object aimed at or as to imply fraud, mistake, circumvention or oppression. There is no sound reason why persons competent and free to contract may not agree upon this subject as fully as upon any other, or why their agreement, when fairly and understandingly entered into with a view to just compensation for the anticipated loss, should not be enforced."

[The court went on to reason] ". . . [T]he later rule, however, is to look with candor, if not with favor, upon such provisions in contracts when deliberately entered into between parties who have equality of opportunity for understanding and insisting upon their rights, as promoting prompt performance of contracts and because adjusting in advance, and amicably, matters in the settlement of which through courts would often involve difficulty, uncertainty, delay and expense . . . ."

". . . It is obvious that the extent of the loss which would result to the Government from delay in performance must be uncertain and difficult to determine and it is clear that the amount stipulated for is not excessive . . . ."

"The parties . . . were much more competent to justly determine what the amount of damage would be, an amount necessarily largely conjectural and resting in estimate, than a court or jury would be, directed to a conclusion, as either must be, after the event, by views and testimony derived from witnesses who would be unusual to a degree if their conclusions were not, in a measure, colored and partisan."

Affirmed [for defendant city of Chicago].

DISCUSSION QUESTIONS

1. Contractor agrees to build an addition to Owner's factory. As the work was getting underway, a sudden flood practically destroyed the factory. What are Owner's options if it is decided not to rebuild the factory?

2. Describe the conditions under which the courts will enforce a contract provision for liquidated damages.

3. When would rescission be an appropriate remedy? Since "restitution" is customarily coupled with rescission, describe that concept.

4. Discuss the theory of the obligation to mitigate damages and explain its application.

5. Discuss the possible remedies where the construction contract specifications for a 10-unit condominium calls for exterior paint to be applied at the rate of 10 millimeter thickness per coat and inspection discloses that the paint was applied at a 9 millimeter thickness. Can the owner (1) eject the contractor from the project and hire a new contractor, (2) seek an injunction of specific performance to repaint the whole complex, or (3) sue for breach of contract to obtain damages; if so, how are the damages measured?

6. In the *Malo case*, does the court convincingly justify the use of parol evidence to establish a contractual condition? Was this a condition precedent or condition subsequent and what difference, if any, does it make?

7. After a breach of the contract for construction of a building by Contractor, to what extent will the court grant a remedy of specific performance to the Owner who desires to have the building completed?

8. Since specialized insurance is common (e.g., travel insurance at airports and accident insurance at schools), how would the courts view a mandatory insurance policy tied into the sale of an automobile covering injuries from design and manufacturing defects?

6

CRIMINAL LAW, PROCEDURE, AND SANCTIONS

6.1 BUSINESS, PROFESSIONAL ACTIVITIES, AND CRIMINAL LAW

Crimes are public wrongful acts prohibited by the sovereign. Criminal prosecutions are brought by a representative (prosecutor) of the sovereign in the name of a state or of the United States, and those who are convicted of committing criminal acts are subject to punishment imposed by society, such as fines, imprisonment, or execution. Although the same act may constitute both a crime and a tort, the crime is an offense against the public pursued by the sovereign, while the tort is a private injury pursued by the injured party. *Fitzgerald v. United States*, 20 S. Ct. 944 (1900). For example, an arsonist is not only criminally responsible but also is liable to the injured victim for the damage caused by the fire in a civil law suit. The reason that victims of crimes only infrequently file civil lawsuits against their offenders is simply that most criminal defendants are financially unable to pay a damage award.

The business community is not only a victim of crimes such as theft, embezzlement, and forgery but business entities may engage in conduct constituting a crime such as false advertising, bid rigging, and destruction of the environment. In fact, a number of specific criminal laws are aimed at the business community with respect to illegal business practices. Violations of the criminal law by business, tarnishing their reputations, are all too frequently headline news. These crimes in which business is either the victim or the actor increase the cost of doing business, and those costs are ultimately borne by the consuming public.

Criminal conduct defined as a compound concept generally results only from the concurrence of an evil-meaning mind ("intent") with an evil-doing hand ("act"). As a result of this intensive individualistic concept that took deep and early root in American soil, a crime results from the concurrence of prohibited conduct and a culpable mental state. *People v. Cotta*, 49 Cal. 166 (1874). So, in the ordinary sense, an evil deed, without the requisite intent, does not rise to the definition of a crime. The subject of criminal intent is discussed in more detail in section 6.3 of this chapter.

6.2 CLASSIFICATION OF CRIMES

Crimes are usually classified either as felonies or as misdemeanors, depending upon the seriousness of the offense.

Felonies

Felonies usually are defined as crimes punishable by either death or imprisonment for 1 year or more. Ordinarily, the procedure for charging an accused with a felony involves the court's binding the accused over to the grand jury based on probable cause. Evidence is then presented to the grand jury and the accused may be indicted if the grand jury finds there is sufficient evidence to return what is called a "true bill."

In some cases, conviction of a felony may result in disenfranchisement (loss of the right to vote) and may bar a person from practicing certain professions.

Misdemeanors

Misdemeanors are lesser crimes and are generally defined as those punishable by fine or imprisonment for less than 1 year or by imprisonment in a place other than a state prison (i.e., confinement in the local jail). Modern statutory definitions of felonies and misdemeanors now vary from state to state.

Conspiracy

At Common Law, the crime of "conspiracy" and "attempt" were described as inchoate (just begun or incomplete) offenses. Conspiracy, described as a partnership in crime, results from an agreement between two or more persons to commit an "unlawful act" or to commit a lawful act by "unlawful means." *People v. McManis*, 266 P.2d 134 (Cal. App. 1954). The societal threat posed by the nature of this conduct resulted from the banding together of two or more persons and a pooling of their talents and resources for a common unlawful purpose. By way of illustration, this crime is frequently committed in situations where two or more persons agree to obstruct justice by suppressing or fabricating evidence.

While the crime of conspiracy refers to an "agreement," no formal or written agreement is necessary, and it is sufficient that there is an understanding between the parties. However, the modern statutory crime of conspiracy generally requires an overt act by one of the parties in addition to the agreement, although the overt act need not itself amount to a crime and may be satisfied by such acts as making a telephone call or mailing a letter.

Proof that the conspirators agreed on the details of their criminal enterprise is not required, but it must be demonstrated that there was agreement on the "essential nature of the plan." In *United States v. Gallishaw*, 428 F.2d 760 (2d Cir. 1970), the conviction of the accused was reversed because of inadequate instructions to the jury on the scope of the conspiratorial agreement. There the accused was alleged to have supplied codefendants with a machine gun that was later used to rob a federally insured bank. According to the testimony when the accused supplied the gun, one of the codefendants said he would either "pull the bank job" or "pull something else" with the gun. The trial judge erred in instructions to the jury that the accused "need not know the object of the conspiracy if he understood that the conspiracy was to do something wrong."

A conspirator may withdraw from the conspiracy, but the withdrawal must be communicated at such a time that would permit the other conspirators to abandon their own effort toward commission of the target offense or any other offense committed by the co-conspirator in pursuance of the common plan. *Loser v. Superior Court of Alameda County*, 177 P.2d 320 (Cal. App. 1947). Under the ALI Model Penal Code Section 5.05(1), the penalty for conspiracy is the same as the penalty provided for the target offense.

Attempt

At Common Law, a person commits the crime of an "attempt" when the accused, with intent to commit a particular crime, performs an act that tends toward but falls short of the consummation of such crime. *People v. Anderson*, 12 Cal. Rptr. 500, 361 P.2d 32, (Cal. 1961), cert. den. 82 S. Ct. 368 (1961). To constitute an attempt, there must be an intent to commit a particular crime although such intent may be inferred by the accused's conduct. However, mere intent to commit a crime is not enough, and there must be performance of an act. Likewise, mere preparation such as buying a gun does not constitute an "attempt" to murder. On the other hand, lying in wait for an intended victim is sufficient. Section 5.01 of the ALI Model Penal Code defines the crime of "attempt" as ". . . essentially one of criminal purpose implemented by an overt act strongly corroborative of such purpose." An attempt, at Common Law, was punishable as a misdemeanor whether the target crime was a felony or a misdemeanor.

6.3 INTENT: CRIMINAL STATE OF MIND

Social Aspects of Crimes

At Common Law, the range of crimes was limited and well defined. Now, a wide range and variety of conduct is defined as criminal. Today, crimes have a statutory basis resulting from legislative enactment of specific criminal statutes defining particular conduct as criminal. The definitions of criminal conduct change with time and are resolved as social and political issues. Some behavior that was considered criminal (e.g., possession of alcoholic beverages and interracial marriages) is no longer characterized as criminal, and, today, there are many proposals to "decriminalize" various kinds of behavior such as gambling, prostitution, consensual sex, and possession of drugs. Those who argue for decriminalization maintain that attempts to treat "victimless" crimes as criminal are ineffective, cause corruption, overburden police work, clutter court calendars, and result in a loss of respect for the law.

General Criminal Intent

The general theory of criminal conduct at Common Law was the combination of an act and criminal intent. Criminal intent was defined as the evil mind, and, although not synonymous with motive, proof of the motive element of the crime may be circumstantial evidence of intent.

"General intent crimes," a residual term, is used to describe any crime that does not require a specific criminal intent. Many criminal statutes now remove the need for the proof of general criminal intent by substituting a conceptual theory that persons intend the natural consequences of their acts, and it may be used to encompass acts as gross negligence (recklessness) or conduct taken in utter disregard for the consequences of such action. By way of illustration, in the crime of manslaughter resulting from the reckless operation of a motor vehicle, the criminal intent is supplied by equating gross negligence to general intent.

Specific Intent

The definition of certain crimes includes an element of specific criminal intent. For example, the crime of burglary is defined as including the elements of (1) breaking and entering, (2) a dwelling, (3) in the night time, and (4) with intent to commit a felony therein. The first three elements are merely matters of factual proof and judicial construction. The last element requires that the prosecution establish that the accused "intended to commit a felony."

Specific intent crimes focus on what the accused was actually thinking or planning at the time of the offense. This element might be satisfied by contemplated conduct, such as in "securities" crimes where the accused "intended to defraud," or by contemplated knowledge of the accused, such as in "knowingly receiving stolen property."

Strict Liability

The modern trend of criminal law has extended the scope of prohibited conduct, especially with respect to business activities to crimes involving "strict liability" or "liability without fault" by imposing culpability without proof of the existence of mens rea (guilty mind).

The landmark case of *United States v. Park*, 421 U.S. 658 (1975), reflects the Supreme Court's current thinking on the issues of (1) the statutory elimination of a need for an awareness of some wrongdoing and (2) responsibility of top corporate personnel. In that case, Park, president of Acme Markets, Inc. (a large national food chain with 36,000 employees, 874 retail outlets, 12 general warehouses, and 4 special warehouses) was convicted of violating the Federal Food, Drug, and Cosmetic Act (FDCA) by receiving food shipped in interstate commerce and storing it in a Baltimore, Maryland, warehouse where it was exposed to contamination by rodents. Park was notified by the federal inspectors of unsanitary conditions in Acme's Baltimore warehouse and in the previous year had been notified of similar conditions at Acme's Philadelphia, Pennsylvania, warehouse. After receiving notice, Park had conferred with Acme's vice president for legal affairs and was told that corrective action had been taken. The following year, a federal follow-up investigation showed some improvement but still showed evidence of rodent contamination, which resulted in the criminal charge.

At the trial, defendant argued "Park was not personally concerned in the violation and while all employees were under his general direction, the company had an organizational structure with responsibilities [for functions assigned] . . . according to different phases of its business . . . and that Park had assigned these duties to individuals who in turn had staffs under them."

In affirming defendant Park's conviction, the Court rejected the defense that subordinates had been assigned to take corrective action and held

> The purpose of the FDCA was to prevent adulteration of foods and drugs which articles touch phases of the lives and health of people which, in the circumstances of modern industrialism, are largely beyond self-protection. The FDCA is now a familiar type of statute *which dispenses with the conventional requirement of criminal conduct*—awareness of some wrongdoing. In the interest of the larger good, the FDCA puts the burden of acting at hazard upon a person otherwise innocent but standing in a responsible relation to a public danger. [The basis of the conviction of Park the President was grounded on] his *responsible relation to the situation and by virtue of his position [and having] authority and responsibility to deal with the situation* [emphasis added].

In the regulatory context, courts have rejected the position that a corporation should be absolved from criminal responsibility where it has made good faith efforts to ensure compliance with the law. Chief Judge Magruder, in *St. Johnsbury Trucking Co. v. United States*, 220 F.2d 393 (1st Cir. 1955), in a concurring opinion noted that with regard to certain Interstate Commerce Commission (ICC) regulations requiring the labeling of motor vehicles and trailers transporting dangerous substances, the corporate defendant could not absolve itself from liability by showing that high executives of the corporation "took utmost care to lay down for the guidance of subordinates employee procedures designed to ensure compliance with the regulations."

Knowingly

The term "knowingly" has been the subject of considerable confusion, and litigation has given that word several shades of meaning. In *U.S. v Schneiderman*, 102 F. Supp. 87, 93 (S.D. Cal. 1951), the court defined "knowingly" as meaning "to act voluntarily and purposely, and not because of mistake or inadvertence or other innocent reason." However, the term does not imply a necessity of an element of evil purpose, but "does import knowledge of the essential facts from which the law presumes the knowledge of the consequences arising therefrom."

6.4 CAPACITY TO COMMIT CRIMES

Ordinarily, persons who have reached the age of criminal responsibility, who are mentally capable, and who act on their own will, that is, not under duress or in self defense, are responsible for the commission of their criminal acts. While criminal intent (general or specific) is ordinarily an element of criminal conduct, intent may be inferred from the nature of the defendant's acts, if the defendant has the capacity to form the required criminal intent. The idea underlying the requirement of intent is based on the philosophy that criminal law only seeks to punish conscious wrongdoers. Criminal law recognizes several kinds of incapacity: infancy, insanity, and intoxication.

Criminal Capacity of Infants

By English Common Law, following the rule of the Roman law, a child under the age of 7 years is conclusively presumed to be incapable of committing a crime since there could be no criminal intent. This rule generally prevails today, although the statutes of some states have raised the age limit. The Common Law also raises a rebuttable presumption of incapacity in favor of an infant between the ages of 7 and 14. Youth between the ages of 14 and 21 were presumed capable of forming criminal intent.

Capacity of an Insane Person

A person cannot be punished for any act committed while insane, although the same act would be criminal if done by a sane person. Since criminal intent is an essential element of crime, a person who is insane is incapable of forming any intent. There are various forms of mental deficiency or derangement that will not excuse commission of a crime, and, to be excused, the person's insanity must have caused the accused to be incapable of forming a criminal intent.

The courts do not agree on the test of insanity that will excuse the commission of a crime. Some jurisdictions apply the so-called "irresistible impulse test," under which the accused will be excused, notwithstanding being able to comprehend the nature and consequences of the act and knowing that it was wrong, if the accused was forced to its execution by an impulse that the accused was powerless to control as a result of a diseased mind. Other jurisdictions adhere to the early English rule established in the *M'Naghten case*, in which the defense of insanity is determined by a test of whether the accused at the time of the commission of the crime was laboring under such a defect of reasoning as not to know the nature and quality of the act and did not know or have the ability to distinguish between right and wrong.

Persons are presumed sane, and the accused must affirmatively prove the defense of insanity. Insanity of the accused may also affect a criminal trial in another manner, where the accused is incapable of assisting in the defense of the case. In that event, the trial may be delayed until the accused regains sanity.

Intoxication of Accused

Voluntary intoxication is generally not a complete defense to criminal liability. It can, in some cases, diminish the extent of a defendant's criminal conduct if it prevents the formation of a specific criminal intent. For example, since first-degree murder under some state statutes requires proof of premeditation (a conscious decision), the accused, who is intoxicated, may not be capable of premeditation. In that case, the accused may be convicted only of second-degree murder, which does not require premeditation. Involuntary intoxication may be a complete defense to criminal liability.

6.5 BURDEN OF PROOF

General

Normally, the plaintiff's burden of proof in a civil action (torts and contracts) is a preponderance of the evidence. This simply means that when both sides have presented their evidence, the plaintiff may recover only if the greater weight of the believable evidence produced was on the plaintiff's side. This standard of proof generally is applied in civil cases where only money is at stake, in contrast to criminal cases where the defendant's life or liberty may be at stake.

For a person to be convicted of criminal behavior, the prosecutor must (1) demonstrate a statutory prohibition existing at the time of the act; (2) prove, beyond a reasonable doubt, that the defendant committed the criminal act; and (3) prove that the defendant had the capacity to form a criminal intent.

Proof Beyond a Reasonable Doubt

In criminal cases, where the life and liberty of the accused person is at stake, the legal system places strong limits on the power of the prosecutor to convict a person of crime. First, criminal defendants are presumed innocent. Second, the state must overcome this presumption of innocence by proving every element of the offense charged against the defendant beyond a reasonable doubt. This means beyond any substantial doubt in fact or in law.

The prosecutor must prove its case within the framework of criminal law procedural safeguards and Constitutional rights that are designed to protect the accused. The prosecutor's failure to prove any material element of its case results in the accused being acquitted, even though the accused may actually have committed the crime charged.

6.6 STATUTORY AND CONSTITUTIONAL SAFEGUARDS AND DEFENSES

Statutory Basis of Crimes

Generally, today there are no Common Law crimes existing in the states, although the courts, at times, may refer to Common Law crimes for the purpose of construing statutory crimes. Before behavior can be treated as criminal, the legislature must have passed a statute making specified conduct criminal. Then, only those who commit the prohibited act after passage of the statute may be prosecuted. The power of Congress and the state legislatures to make behavior criminal is constitutionally limited in two ways. First, behavior that is protected by either the U.S. Constitution or a state constitution cannot be made criminal. For example, the exercise of free speech protected by the First Amendment to the Constitution cannot be made criminal. Second, criminal statutes must also define the prohibited behavior clearly, so that an ordinary person would understand what behavior is prohibited. This requirement comes from the Due Process Clauses in the Fifth and Fourteenth amendments to the U.S. Constitution.

Constitutional Protections

In addition to the presumption of innocence, the U.S. legal system has built-in several other safeguards to protect the accused. These safeguards are designed to prevent innocent people from being convicted for crimes they did not commit and also to represent an ideal of the proper role of government in a democracy. As Supreme Court Justice Oliver Wendell Holmes remarked, "I think it less evil that some criminals should escape than that the government should play an ignoble part."

Some of the safeguards enjoyed by the accused in our legal system follow:

1. Defendants who have been acquitted of a crime may not be tried again for the same offense by the same jurisdiction. This is known as the prohibition against double jeopardy.

2. Defendants in criminal cases have a right to remain silent and cannot be compelled to testify against themselves.

3. Illegally obtained evidence, resulting from "unreasonable searches and seizures," is prohibited by the Fourth Amendment from being used by the state in criminal prosecutions.

4. Persons charged with crimes have a right to be represented by an attorney if imprisonment is a possible result of conviction.

5. Persons accused of crimes have the right to confront and cross-examine their accusers.

6. When the police take a person into custody, they must advise the person of the right to remain silent and the right to counsel. Confessions made in the absence of these warnings are inadmissible as evidence in court and cannot be used to convict a person.

Against this backdrop, the Supreme Court opinion in *Dow Chemical Co. v. United States*, 749 F.2d 307, (6th Cir. 1984), aff'd. 106 S. Ct. 1819 (1986) presents interesting questions concerning unconstitutional search. In that case, the Environmental Protection Agency (EPA) had made an earlier on-site inspection of the Dow plant to determine compliance with smoke stack emissions standards from two of the company's power plants. Because of the company's refusal to permit subsequent on-site inspections without an appropriate search warrant, EPA overflew the company's plant on six occasions at altitudes of between 1,200 and 20,000 feet to take photographs. These photos, which show the company's 2,000 acre Midland, Michigan, complex, consisting of a number of covered buildings and some outdoor manufacturing areas, were enlarged to the scale that 1 inch was equivalent to 20 feet. These enlargements were sent to the EPA Chicago Regional Enforcement Office.

In pitting the company's constitutional rights and business secret interests against the government's need to enforce the environmental statutes, the Court came down on the side of the EPA by holding that the "aerial surveillance did not violate Dow's Fourth Amendment rights" (based on a designation of the plant), because of its sprawling size and make up, as an "open field." The Court having made that determination concluded that the constitutional protection to the people in their "persons, houses, papers, and effects" does not extend to open fields. The Court's analysis was limited to the question of whether the Dow plant was an "open field" and did not consider a number of other factors bearing on the privacy rights of a business such as the height of the surveillance, its duration, what methods of observation enhancement were used, and if competitors could have access to the photographic information.

Former Jeopardy

The Fifth Amendment to the U.S. Constitution and comparable provisions in state constitutions protect individuals by providing that "no person shall be subject for the same offense to be twice put in jeopardy of life and limb." *United States v. Benz*, 51 S. Ct. 113 (1931), *Ex Parte McLaughlin*, 41 Cal. 211 (1871). This federal guarantee is also binding upon the states through the Fourteenth Amendment. This guarantee rests on the reasoning that the resources and power of the government should not be allowed to make repeated attempts to convict an individual for an alleged offense and subject the accused to continued embarrassment and expense.

Entrapment

Entrapment is a defense commonly raised in certain crimes, such as illegal sale of drugs, and refers to the fact that the criminal intent did not originate with the accused but originated with the police or other law enforcement personnel. Under those circumstances, when a criminal act is committed at the instigation of government, the prosecution should not be able to contend that the accused is guilty of the crime. While the government may provide the opportunity for the accused to commit the criminal act (e.g., the purchase of drugs by undercover agents), the accused must have voluntarily intended to engage in criminal conduct absent the undercover agent's action. Because entrapment, as an affirmative defense, must be proved by the accused, the nature of the defense requires that the accused admit the act was committed in order to argue that the police were the motivating cause in the commission of the criminal act.

6.7   WHITE COLLAR CRIMES

People in business today are more likely than ever before to have some unpleasant contact with the criminal justice system. Business may not always be the victim but rather may be the active criminal. Business crimes, known as "white collar crimes," usually involve fraudulent non-violent conduct, as contrasted with the popular perception of "street crimes." There is a trend today to make violations of regulatory statutes criminal offenses punishable by fines and/or imprisonment. Many prosecutors and judges are demonstrating a "get tough" attitude toward white collar crimes that in the past have often been treated leniently. The argument supporting this harsher approach has been advanced that personal liability including jail and prison sentences for business executives is necessary to deter business from violating laws, since the tendency is to view any fines imposed on the company as merely a cost of doing business.

This attitude on the part of government officials probably results from several factors. There has been a long-standing public outcry against the lenient treatment of white collar crime. This complaint has been aggravated by statistics indicating the tremendous cost of such crime (amounting to billions of dollars, annually) and reflects the post-Watergate atmosphere of public hostility toward criminal conduct by persons in positions of power and authority. In addition to seeking jail sentences for business defendants, prosecutors are also using a range of criminal statutes that include mail fraud and conspiracy to indict and convict them.

While a number of people have written in the area of white collar crime, sociologist Edwin H. Sutherland's *White Collar Crime*, (Yale University Press, New Haven, Conn., 1983) presents a scholarly and comprehensive sociological review of business attitudes and conduct. In criticizing criminal statistics, Sutherland insists that "conviction is important from the point of view of the authority of public agencies to administer punishment [but] is not important as a definition of criminal behavior." In responding to the question of whether some action is white collar crime, it was noted that from such conduct persons injured may be divided into two groups: first, a relatively small number of persons engaged in the same occupation or directly affected, and, second, the general public either as consumers or as constituents of the general social institutions affected by the violations. Sunderland's studies focused on four different common business crimes: (1) the anti-trust laws, (2) false and deceptive advertising, (3) coercive unfair labor practices, and (4) infringements of patents, copyrights, and trademarks. The anti-trust and deceptive advertising laws were designed to shield competitors and the system of free competition from rigged bids, unfair competition, and fraud. The labor laws were to protect the employees from coercion and the public from interferences by strikes and boycotts. Crimes relating to patent rights and trade secrets were to promote progress in science by preventing theft of inventions and other intellectual property.

A list of statutes that are frequently the subject of white collar crimes includes the Securities Act, Trust Indenture Act, Investment Advisors Act, False Statement in Tax Return and Tax Evasion, General False Statement to Federal Employee (Section 1001), Commercial Bribe, Labor Bribe, Political Bribe, Foreign Corrupt Practices Act, Computer Crimes, Environmental Offenses, Occupational Safety and Health Offenses, Wire Fraud, Obstruction of Justice, and Racketeer Influenced and Corrupt Organizations Act.

Particularly important and frequently used is the federal crime of knowingly and willfully making a false statement or material omission, or an intentional or reckless misrepresentation, in any manner within the jurisdiction of a federal agency or department (18 U.S.C. Section 1001 (1976)). This widely used criminal statute requires proof of knowingly and willfully making a material false statement or an intentional misrepresentation and is punishable with a penalty of up to a $10,000 fine or imprisonment of up to 5 years. In *United States v. Matanky*, 482 F.2d 1319, 1322 (9th Cir. 1973) cert. den. 414 U. S. 1039 (1973), the court held that a doctor knowingly submitting false medicare payment vouchers to a private contractor (insurance carrier) working on behalf of a federal agency violated Section 1001.

PERRIN v. UNITED STATES
100 S. Ct. 311, (1979)

Mr. Chief Justice Burger delivered the opinion of the Court.

We granted certiorari to resolve a circuit conflict on whether commercial bribery of private employees prohibited by a state criminal statute constitutes "bribery . . . in violation of the laws of the State in which committed" within the meaning of the Travel Act 18 U.S.C. Section 1952.

Petitioner, Vincent Perrin and four codefendants, were indicted in the Eastern District of Louisiana for violating the Travel Act. The Travel Act provides in part:

(a) Whoever travels in interstate or foreign commerce or uses any facility in interstate or foreign commerce including the mail, with intent to --

(1) distribute the proceeds of any unlawful activity; or
(2) commit any crime of violence to further any unlawful activity;

(b) As used in this subsection "unlawful activity" means (1) any business enterprise involving gambling . . . controlled substances . . . or prostitution offenses in violation of the laws of the State in which they are committed or of the United States, or (2) extortion, bribery, or arson in violation of the laws of the State in which committed or of the United States.

The indictment charged that Perrin and his codefendants used the facilities of interstate commerce for the purpose of promoting commercial bribery schemes in violation of the laws of the State of Louisiana. Following a jury trial, Perrin was convicted.

The Government's evidence at trial was that Perrin, David Levy, and Duffy LaFont engaged in a scheme to exploit geological data obtained from the Petty-Ray Geophysical Company. Petty-Ray, a Louisiana-based company, was in the business of conducting geological explorations and selling the data to oil companies. At trial, company executives testified that confidentiality was imperative in the conduct of their business. The economic value of exploration data would be undermined if its confidentiality were not protected.

In June 1975 LaFont importuned Roser Willis, an employee of Petty-Ray, to steal confidential geological exploration data from his employer. In exchange, LaFont promised Willis a percentage of the profits of a corporation which had been created to exploit the stolen information. Willis' position as an analyst of seismic data gave him access to the relevant material, which he in turn surreptitiously provided to the conspirators. Perrin, a consulting geologist, was brought into the scheme to interpret and analyze the data. In July 1975 Perrin met with Willis, LaFont, and Levy. Perrin directed Willis to call a firm in Richmond, Texas, to obtain gravity maps to aid him in his evaluation. After the meeting, Willis contacted the Federal Bureau of Investigation and disclosed the details of the scheme.

The United States Court of Appeals for the Fifth Circuit affirmed Perrin's conviction, rejecting his contention that Congress intended "bribery" in the Act to include only bribery of public officials.

Petitioner argues that Congress intended "bribery" in the Travel Act to be confined to its common law definition, i.e., bribery of a public official. He contends that because commercial bribery was not an offense at common law, the indictment fails to charge a federal offense.

The Travel Act was one of several bills enacted into law by the 87th Congress as part of the Attorney General's 1961 legislative program directed against "organized crime." Then Attorney General Robert Kennedy testified at Senate and House hearings that federal legislation was needed to aid state and local governments which were no longer able to cope with the increasingly complex and interstate nature of large scale, multiparty crime.

To remedy a gap in the authority of federal investigatory agencies, Congress employed its now familiar power under the Commerce Clause of the federal Constitution to prohibit activities of traditional state and local concern that also have an interstate nexus. See, e.g., 18 U.S.C. Section 1201 (federal kidnaping statute): 18 U.S.C. Section 2312 (interstate transportation of stolen automobiles). That Congress was consciously linking the enforcement powers and resources of the federal and state governments to deal with traditional state crimes is shown by its definition of "unlawful activity" as an "enterprise involving gambling, liquor . . . narcotics or controlled substances . . . or prostitution offenses in violation of the laws of the State in which they are committed or of the United States." The statute also makes it a federal offense to travel or use a facility in interstate commerce to commit "extortion [or] bribery . . . in violation of the laws of the State in which committed or of the United States." Because the offenses are defined by reference to existing state as well as federal law, it is clear beyond doubt that Congress intended to add a second layer of enforcement supplementing what it found to be inadequate state authority and state enforcement.

A fundamental canon of statutory construction is that unless otherwise defined, words will be interpreted as taking ordinary, contemporary, common meaning. Therefore, we look to the ordinary meaning of the term "bribery" at the time Congress enacted the statute in 1961. In light of Perrin's contentions we consider first the development and evolution of the common-law definition.

At early common law, the crime of bribery extended only to the corruption of judges. The writings of a 19th century scholar inform us that by that time the crime of bribery had been expanded to include the corruption of any public official and the bribery of voters and witnesses as well. And by the 20th century, England had adopted the Prevention of Corruption Act making criminal the commercial bribery of agents and common carrier and telegraph company employees, labor officials, bank employees, and participants in sporting events.

A similar enlargement of the term beyond its common-law definition manifested itself in the States prior to 1961. Fourteen states had statutes which outlawed commercial bribery generally. An additional 28 had adopted more narrow statutes outlawing corrupt payments to influence private duties in particular fields, including bribery of agents, common carriers and telegraph company employees, labor officials, bank employees, and participants in sporting events.

In sum, by 1961 the common understanding and meaning of "bribery" had extended beyond its early common-law definitions. In 42 states and in federal legislation, "bribery" included the bribery of individuals acting in a private capacity. It was against this background that the Travel Act was passed.

On a previous occasion we took note of the sparse legislative history of the Travel Act. The record of the hearings and floor debates discloses that Congress made no attempt to define the statutory term "bribery," but relied on the accepted contemporary meaning. There are ample references to the bribery of state and local officials, but there is no indication that Congress intended to so limit its meaning. Indeed, references in the legislative history to the purposes and scope of the Travel Act, as well as other bills under consideration by Congress as part of the package of "organized crime" legislation aimed at supplementing state enforcement, indicate that Members, Committees, and draftsmen used "bribery" to include payments to private individuals to influence their actions.

Petitioner also contends that commercial bribery is a "management," or "white-collar" offense not generally associated with organized criminal activities. From this, he argues that Congress could not have intended to encompass commercial bribery [within the Travel Act].

The notion that bribery of private persons is unrelated or unknown to what is called "organized crime" has no foundation. The hearings on the Travel Act make clear that a major area of Congressional concern was with the infiltration by organized crime into legitimate activities. Indeed, the McClellan Committee in 1960, like the Kefauver Committee in 1950-1951, documented numerous specific instances of the use of commercial bribery by these organized groups to control legitimate businesses.

There can be little doubt that Congress recognized in 1961 that bribery of private persons was widely used in highly organized criminal efforts to infiltrate and gain control of legitimate businesses, an area of special concern of Congress in enacting the Travel Act.

An [earlier] opinion by Chief Justice Earl Warren for a unanimous Court rejected the argument limiting the definition of extortion to its common-law meaning, holding that Congress used the term in a generic and contemporary sense. The Court noted that in 1961 the Attorney General had pressed Congress to include "shakedown rackets," "shylocking," and labor extortion, which were methods frequently used by organized groups to generate income and infiltrate legitimate activities.

Congress has clearly stated its intention to include violations of state as well as federal bribery law. Until statutes such as the Travel Act contravene some provision of the Constitution, the choice is for

Congress, not the courts. We hold that Congress intended "bribery . . . in violation of the laws of the State in which committed" as used in the Travel Act to encompass conduct in violation of State commercial bribery statutes. Accordingly, the judgment of the Court of Appeals is affirmed.

UNITED STATES v. FREZZO BROTHERS, INC.
602 F.2d 1123 (3rd Cir. 1979)

Rosenn, Circuit Judge. Since the enactment in 1948 of the Federal Water Pollution Control Act (FWPCA), the Government has, until recent years, generally enforced its provision to control water pollution through the application of civil restraint. In this case, however, the Government in the first instance has sought enforcement of the Act as amended in 1942 against an alleged corporate offender and it's officers by criminal sanctions.

The appellants were convicted by a jury on six counts of willfully or negligently discharging pollutants into a navigable water of the United States without a permit, in violation of [Section 301 and 309(c)]. The corporate defendant, Frezzo Brothers, Inc., was fined $50,000 and the individual defendants, Guido and James Frezzo, received jail sentences of 30 days each and fines aggregating $50,000. The Frezzos appeal from the trial court's final judgment of sentence. We affirm.

Frezzo Brothers, Inc., is a Pennsylvania corporation engaged in the mushroom farming business near Avondale, Pennsylvania. The business is family operated with Guido and James Frezzo serving as the principal corporate officers. As a part of the mushroom farming business Frezzo Brothers, Inc., produces compost to provide a growing base for the mushrooms. The compost is comprised mainly of hay and horse manure mixed with water and allowed to ferment outside on wharves.

The Frezzos' farm had a 114,000 gallon concrete holding tank designed to contain water run off from the compost wharves and to recycle water back to them. The farm had a separate storm water run-off system that carried rain water through a pipe to a channel box located on an adjoining property owned by another mushroom farm. The channel box was connected by a pipe with the unnamed tributary of the East Branch of the White Clay Creek. The waters of the tributary flowed directly into the creek.

Counts One through Four of the indictment charged the defendants with discharging pollutants into the East Branch of the White Clay Creek in [July and September 1977]. On these dates Richard Casson, a Chester County Health Department investigator, observed pollution in the tributary flowing into the Creek and collected samples of wastes flowing into the channel box. The wastes had the distinctive characteristics of manure and quantitative analysis of the samples revealed a concentration of pollutants in the water. The Government introduced meteorological evidence at trial showing that no rain had been recorded in the area on these [dates]. Based on this evidence the Government contended that the Frezzos had willfully discharged manure into the storm water run-off system that flowed into the channel box and into the stream.

[Investigator Casson returned to the Frezzo farm where he took water samples and photographed the facilities.] The Government theorized that the holding tank was too small to contain the compost wastes [under certain conditions].

[The Frezzos contend] that the indictment should have been dismissed because the EPA had not promulgated an effluent standard applicable to the compost manufacturing business. The Frezzos argue that before a violation of [the FWCPA] can occur, the defendants must be shown to have not complied with the existing effluent limitations under the Act. The District Court disagreed, finding no such requirement.

We see nothing impermissible with allowing the Government to enforce the Act by invoking [Section 301] even if no effluent limitations have been promulgated for the particular business charged with polluting. Without this flexibility numerous industries not yet considered as serious threats to the environment may escape administrative, civil, or criminal sanctions merely because the EPA has not established the effluent limitations. Thus, dangerous pollutants could be continually injected into the water solely because the administrative process has not yet had the opportunity to fix specific pollution limitations. Such a result would be inconsistent with the policy of the Act.

[I]f no effluent limitations have yet been applied to an industry, a potential transgressor should apply for a permit to discharge pollutants under Section 402(a). The pendency of a permit application, in an appropriate case, should shield the applicant from liability for discharge in the absence of a permit.

In the present case, it is undisputed that there was no pending permit to discharge pollutants; nor had Frezzo Brothers, Inc., ever applied for one. This case, therefore, appears to be particularly compelling for broad enforcement under Sections 301(a), 309(c)(1). The Frezzos, under their interpretation of the statute could conceivably have continued polluting until EPA promulgated effluent limitations for the compost operation. The Government's intervention by way of criminal indictment brought to a halt potentially serious damage to the stream in question, and as no doubt alerted EPA to the pollution problems posed by compost production. We therefore hold that the promulgation of effluent limitations standards is not a prerequisite to the maintenance of a criminal proceeding based on violation of Section 301(a) of the Act.

The Frezzos maintain that the Government on the evidence failed to establish a willful act. We disagree. The jury was entitled to infer from the totality of the circumstances surrounding the discharges that a willful act precipitated them. The Government did not have to present evidence of someone turning on a valve or diverting waste in order to establish a willful violation of the Act.

UNITED STATES v. BAKER
626 F.2d 512 (5th Cir. 1980)

Goldberg, Circuit Judge. Appellants Baker and Knowlton were part-time employees of the Housing Authority of the City of Dallas (DHA). hired to help train DHA security guards. DHA was the recipient of $3,500,000 from the United States Department of Housing and Urban Development (HUD) to improve conditions at low-income housing projects operated by DHA. On a quarterly basis, DHA was required to report to HUD the ways in which the federal money was being spent.

Baker and Knowlton were convicted under 18 U.S.C. Section 1001 of submitting to DHA false time sheets which claimed pay for hours not actually worked. The questions presented on appeal are whether the evidence at trial was sufficient to sustain a finding that the false statements made by the appellants were material, and whether the trial court properly instructed the jury with regard to defendants' theory of the case. We answer both questions in the affirmative, and we affirm.

[P]roof of five elements is essential to sustain a conviction under the false statement proscription of Section 1001: (1) a statement, (2) falsity, (3) materiality, (4) specific intent, and (5) agency jurisdiction. In the case at hand, appellants Baker and Knowlton challenge the trial court's finding of the third of these elements, and argue that the false statements made were not material because they could not affect any decision of HUD, the federal agency involved. Since the federal funds had already been appropriated at the time the false statements were made, and since Baker and Knowlton were paid not by HUD but by the

DHA, appellants suggest that HUD could not have been influenced by the submission of the false time sheets.

The purpose of the materiality requirement of 18 U.S.C. Section 1001 is "to exclude trivial falsehoods from the purview of the statute." While the requirement of materiality explicitly applies only to the first clause of Section 1001, "courts have inferred a judge-made limitation of materiality" on the second ("false statements") clause as well. . . . In addition, it is well-settled that a false statement need not be made directly to a federal agency to sustain a Section 1001 conviction as long as federal funds are involved. . . .

In the case at hand, the statements made by appellants resulted in their each receiving substantial wages to which they were not entitled. The performance of the program was hindered since the federal funds were exhausted more swiftly than they otherwise would have been. In addition, the quarterly DHA reports required by HUD evidence the fact that the federal agency had "the ultimate authority to see that the federal funds [were] properly spent."

The false statements were thus material in that they had a substantial effect on the federal funds forwarded by HUD, on the performance of the DHA project, and on the pay received by the appellants.

DISCUSSION QUESTIONS

1. Should economic infeasibility be a defense to a criminal charge involving either a jail sentence or imprisonment?

2. With respect to charges under Section 1001 with respect to "false statements to law enforcement agents," is it a defense that the federal agent knew that the statement was false at the time? (see, *U.S. v. Johnson*, 530 F.2d 52 (5th Cir. 1976))? What is the significance of statements made to another person (FBI agent) who the accused did not know to be a government agent at the time the false statement was made (see, *U.S. v. Yermain*, 104 S. Ct. 2936 (1984)); suppose the false statements are outside of the jurisdiction of the agency (see, *U.S. v. DiFonzo*, 603 F.2d 1260 (7th Cir. 1979))?

3. Considering the Court's decision in the *Dow case*, would the Court have reached a different decision had the "company" been engaged in raising, processing, and distributing some drug such as marijuana? Did the Court improperly fail to consider such factors and the possibility that "trade secrets" may be in jeopardy as the result of this surveillance?

4. In applying the teaching of the *Park case*, describe some other business activities where "strict liability" or the absence of "wrongdoing" should be ap-

plied? Does this suggest any trend in the law to hold that those who are involved in designing or constructing buildings or products for public use should also face criminal penalties where inadequate review of calculations or design result in injuries?

5. What policy would be most effective in obtaining the maximum compliance with law by business enterprises? Should there be a different policy for small and large corporations, for example, those with 100 employees as contracted with 10,000 employees?

6. In 1970, Congress sought to strike at organized crime's infiltration of legimate business by passage of the Racketeer Influenced and Corrupt Organizations Act (RICO). That Act prohibits from interstate commerce anyone engaged in a "pattern of racketeering" which is defined as the commission of at least two acts within 10 years of any of a variety of already existing crimes including mail and wire fraud. Use of funds from those crimes even in legitimate business activities are covered by the Act. As the result of language in the Act instructing the courts to construe the language broadly "to effectuate the Act's remedial purposes," a number of individuals who would not generally fall within the common meaning of the organized crime classification, have been convicted under the Act. Thus, a number of executives and businesses have been indicted under RICO. What correlations and distinctions are there between the rules of construction applicable to contracts and those applicable to interpretation of criminal statutes?

7

UNINTENTIONAL TORTS
Negligent Conduct
and Duty of Care

7.1 INTRODUCTION

The term "tort" is of Latin derivation meaning "twisted" and comes from the twisting of the body in a form of torture. Tortious conduct involves the failure to observe a standard of care required by civil law as opposed to wrongful conduct under the criminal law. Conduct that amounts to a tort encompasses a wide variety of action, and torts with respect to the person include libel, false imprisonment, and assault and battery; torts involving property include trespass, conversion, and nuisance.

Certain action by a person interfering with the property of another may constitute both a crime (a wrong against society) and a tort (a wrong against a person). The California Code provides that "every person is bound, without contract, to abstain from injuring the person or property of another or infringing upon any of [his or her] rights." Traditionally, torts against the person such as libel involves false and malicious written publication (i.e., communication to another) tending to expose the injured person to ridicule, contempt, or hatred. Traditional torts against property include trespass to another's land such as the extension of fences or other structures that project over property lines or the invasion of another's property by dust, odors, or other pollution.

The so-called Industrial Revolution that began in the early part of the 19th century created a new conceptual question for the law of torts.

Traditional tort injury resulted from intentional conduct such as assault and battery, trespass, or libel. A growing number of injuries to people and property caused by the advent of railroads, machinery, and newly developed tools simply did not fit within the framework of traditional intentional torts but resulted from unintended negligent conduct. The legal system was forced to evolve a new set of rules to deal with these situations, resulting in the emergence of legal concepts of negligence within the scope of tortious conduct.

The underlying concept of torts was originally theoretically grounded on an assessment of the defendant wrongdoer's "fault." While early tort law looked to intentional conduct, the quantum of "blameworthiness" was diminished by subsequent reliance on mere carelessness (negligent conduct) of the defendant. More recently, the law trended toward "strict liability" in certain types of injuries (e.g., food adulteration and product liability) by establishing responsibity despite blameless conduct by the defendant.

In the context of engineering, the events in which tort responsibility may be encountered are numerous. During the design process of either products or buildings, there are many possibilities of error in selecting materials, calculating loads, determining stress factors, or protecting the environment. These errors (negligence) may cause injury to persons or property during the manufacturing or construction process and after completion, users may be injured because of unintentional defective design, poor workmanship, or improper materials. The focus of this chapter is on torts resulting from negligent rather than intentional conduct.

Basically, negligence is unintentional careless conduct representing a middle ground of blameworthiness, which translates legally into breach of duty by the defendant resulting in harm to another. To recover in a lawsuit alleging negligence, plaintiff must prove (1) that the defendant had a duty (not to injure) owed the plaintiff, (2) that the defendant breached that duty, and (3) that the defendant's breach of duty was the actual and legal (proximate) cause of the plaintiff's injuries. To be successful, the plaintiff must also overcome any applicable defenses to negligence liability such as contributory negligence or assumption of risk that may be raised by the defendant.

7.2 STANDARD OF CONDUCT: ELEMENTS OF CAUSE OF ACTION

Duty and Breach: Foreseeability

This area of tort law deals with the duty of each member of society not to harm the person or property of another and the consequences of a breach of that duty constituting a twisting of normal relationships.

Liability for negligence is not generally imposed unless there exists a duty of due care owed by the defendant to the person injured or the class of persons of which the plaintiff is a member. Stated differently, the duty arises when the plaintiff should have reasonably foreseen that injury would result from the action undertaken. In *Tarasoff v. Regents of University of California*, 17 Cal.3d 425, 434, 551 P.2d 334 (Cal. 1976), the regents and the doctor (defendants), knowing of the patient's mental condition, failed to warn others resulting in the patient's killing a member of the patient's family. The court held that the plaintiffs were within the zone of persons whom the defendants owed a duty to warn. The scope of the defendant's liability includes risks of injury even though (1) another person (the mentally ill person) was the direct cause of the injury, (2) the relationship between the defendant and the plaintiff was remote, and (3) the defendant would not necessarily have foreseen the particular injury that resulted from the negligent conduct.

Just as the facts of the *Tarasoff case* illustrate the establishment of a duty owed the plaintiff, the test to determine the existence of a legal duty and whether the defendant's conduct constituted a breach of that duty is determined in relation to the circumstances of time, place and the persons involved. *United States Liability Insurance Company v. Haidinger-Hayes, Inc.*, 1 Cal.3rd 586, 594, 83 Cal. Rptr. 418, 422 (1970).

Causation

The RESTATEMENT (Second) of Torts in Sections 430-431 (1965) provides "[c]ausation is established if plaintiff shows that defendant's negligent act in some way contributed to the injury; or that, but for the defendant's negligent act, plaintiff's injury would not have occurred." Determining whether the act or omission was the proximate cause of an accident or injury is usually a question of fact. Since injury involving defendant's negligence is a threshold necessity, the "but for" reasoning is questionable circular analysis. The plaintiff normally has the burden of establishing by direct or circumstantial evidence that it was more probable than not that the defendant's conduct was the legal cause of the injury. *Valdez v. J. D. Diffenbaugh Co.*, 51 Cal. App. 3rd 494, 509, 124 Cal. Rptr. 467, 477 (1975). However, in certain cases, the proof may be affected by various presumptions as described in section 7.3 of this chapter.

An intervening act or force that is independent from the defendant's negligence raises the issue of whether the chain of causation has been broken thus relieving the original actor from liability. *Schrimsher v. Bryson*, 58 Cal. App.3rd 660, 664, 127 362 P.2d 345 (Cal. App. 1976). The RESTATEMENT Sections 440-442 read, "[t]he fact that an intervening act is negligent does not necessarily make [the intervening negligent act] a superseding cause; the original actor remains concurrently liable if a reasonable person would not regard it as highly extraordinary that the intervening actor so acted, or if the act is a normal response to a situation created by defendant's conduct." In *Stewart v. Cox*, 13 Cal. Rptr. 521, 524 (1961), the court held that the failure of a third person to protect the plaintiff from harm was not a superseding cause. Further, the defendant is not relieved for negligence, which is a substantial cause of the injury, by intervening acts that plaintiff could have reasonably foreseen at the time of defendant's negligence. In the famous old "Squib case," it was held that if X throws a bomb at Y and Y throws it toward Z to get rid of it, X and not Y is liable for Z's injury. *Scott v. Shepard*, 2 Blackstone 892.

Even if the defendant has breached a duty owed to the plaintiff, the evidence must establish that plaintiff's injury was caused by that breach. Courts often express this idea as "[t]here is no liability for negligence in the air." For example, when defendant speeds down the street, breaching a duty to those in the area, there is no liability to plaintiff who at the same time and on the same block falls down the front steps of his or her house breaking a leg. Although the defendant was negligent, there is no causal connection between that breach of duty and plaintiff's injury and defendant is not liable for plaintiff's injury.

On the other hand, in unintended "ordinary negligence," the general rule holds the defendant liable only for injuries that fall within the scope of the foreseeable risk. If the defendant could have reasonably foreseen some injury to the plaintiff, then the defendant is liable for every injury to the plaintiff that directly results from that negligence. This means that if some physical peculiarity of plaintiff aggravates plaintiff's injuries, the defendant is liable for the full extent of the injuries. For example, plaintiff's head strikes the windshield of plaintiff's car when defendant's negligently driven truck runs into plaintiff. If plaintiff becomes permanently paralyzed owing to an abnormally thin skull although such injury would have only slightly injured a normal person, defendant nevertheless is liable for plaintiff's paralysis. Likewise, negligent persons are generally held liable for diseases victims contract while weakened by their injuries and in some cases for negligent medical care victims receive for their injury.

Negligent persons are also generally liable for injuries sustained by those who are injured while making reasonable attempts to avoid being injured by defendant's negligent acts. So, defendant will be responsible for plaintiff's injuries sustained as a result of plaintiff's diving out of the path of defendant's negligently driven car. It is commonly said that "negligence invites rescue" and that negligent persons should be liable to those who are injured while making a reasonable attempt to avoid injury when endangered by defendant's negligence.

In some cases, a person's tortious conduct may be the cause in fact—the actual or direct—of an incredible series of losses to numerous people. In "intentional" tort cases, as opposed to "unintentional" torts, the courts have traditionally held people liable for all the consequences that directly result from their intentional wrongdoing, however bizarre and unforeseeable they may be.

Special Standards and Duties

The policy underlying the law of negligence is that members of society have a duty to conduct their affairs in a way that avoids injury to others. Requisite behavior is held to an objective standard of conduct described as "a reasonable person of ordinary prudence in similar circumstances." This standard is flexible, since it allows consideration of all the circumstances surrounding a particular accident. Objectivity flows from the use of a "reasonable person" representing a hypothetical norm, not an "average individual." The court's use of the community average criteria for professionals rather than a higher reasonable prudent professional criteria has been criticized, but, in technical or scientific cases, the standard of care for professionals is more or less uniformly described as that "exercised by reputable members of the same profession practicing in the same or a similar locality under similar circumstances and using reasonable diligence and best judgment in the exercise of professional skills and applicable learning." *Clark v. City of Seward*, 659 P.2d 1227 (Alaska 1983).

Normally, a child of immature years will not be held to an adult standard of conduct but must use that degree of care ordinarily exercised by children of like age, intelligence, and experience. *Cummings v. Los Angeles*, 56 Cal. App.2d 258, 263, 14 Cal. Rptr. 668, 671, 263 P.2d 900 (Cal. App. 1961). But a child is held to an adult standard when engaged in an adult activity such as driving an automobile.

In *City of Mounds View v. Walijarvi*, 263 N.W.2d 420, 424 (Minn. 1978), the court addresses the standard of care for professional engineers:

> Because of the inescapable possibility of error which inheres in these services, the law has traditionally required, not perfect results, but rather the exercise of that skill and judgment which can reasonably be expected from similarly situated professions.

To supply proof, the professional engineer would ordinarily use an expert to testify concerning the required standard of care that was allegedly breached. This is especially true where the judge instructs the jury concerning the duty that the plaintiff owes the defendant under the law. The jury then is called upon to decide whether the defendant met that standard considering all the facts of the case. For example, expert evidence is essential where the jury would be required to have technical information beyond the understanding of an ordinary juror to determine whether a mechanical engineering plan for heating and ventilation was negligently designed. *Seaman Unified School District v. Casson Construction Co.*, 594 P.2d 241, (Kans. 1979). The use of expert testimony and witnesses is discussed in detail in chapter 15.

Degrees of Care Owed Particular Classes of Plaintiffs

A variety of people may be called upon to enter construction sites and business establishments under different situations, and the law establishes different degrees of responsibility for the owner or person in control of the premises with respect to trespassers, licensees and invitees.

Trespasser. As a general rule, an owner or the person in control of premises owes no duty to a trespasser other than that of refraining from inflicting willful (intentional) injury. However, many courts recognize an exception and hold the owner or control person under a duty to notify trespassers of known perils. In addition, many jurisdictions recognize the "attractive nuisance" doctrine. Under that doctrine, an owner or control person, maintaining premises in a condition that would be dangerous to children of tender years and such premises would reasonably be expected to attract children, is under a duty to exercise reasonable care to protect trespassing children. This rule, referred to as the "attractive nuisance" doctrine, is an exception to the general rule concerning the duty owed to trespassers.

Licensee. With respect to a licensee (an uninvited person who is permitted on the premises for his own convenience, benefit or pleasure), the owner or control person owes a duty to refrain from active negligence. This degree of negligence is sometimes characterized as "gross negligence."

Invitee. A higher degree of care—that of ordinary care—is owed by the owner or control person to individuals who are directly or impliedly invited to enter the premises. The owner or control person thus must have the premises in such condition as is reasonably safe for the invitee. However, the law does not hold the owner or control person to the standard of an insurer. It is sometimes difficult to distinguish between the licensee and the invitee, and the *Arthur v. Standard Engineering Company case* at the end of this chapter is useful for examining that line of demarcation.

Negligence is also categorized in some jurisdictions as (1) ordinary, (2) gross, and (3) wanton and willful. In *Altman v. Aronson*, 121 N.E. 505 (Mass. 1919), the court explained,

> [n]egligence, without qualification and in its ordinary sense, is the failure of a responsible person, either by omission or by action, to exercise that degree of care, vigilance and forethought which, in the discharge of the duty then resting on him, the person of ordinary caution and prudence ought to exercise under the particular circumstances. It is a want of diligence commensurate with the requirement of the duty at the moment imposed by the law. Gross negligence is substantially and appreciably higher in magnitude than ordinary negligence. It is an act or omission respecting legal duty of an aggravated character as distinguished from a mere failure to exercise ordinary care. It is very great negligence, or the absence of slight diligence, or the want of even scant care. It is heedless and palpable violation of legal duty respecting the rights of others. But it is something less than the willful, wanton and reckless conduct which renders a defendant who has injured another liable to the latter even though guilty of contributory negligence. It falls short of being such reckless disregard of probable consequences as is equivalent to a willful and intentional wrong. Ordinary and gross negligence differ in degree of inattention, while both differ in kind from willful and intentional conduct which is or ought to be known to have a tendency to injure. This definition does not possess the exactness of a mathematical demonstration, but it is what the law now affords.

Business Torts and Other Trends

In recent years, a new area of torts emerged in importance and is now classified under the title "business torts." Professor William Prosser, a leading authority on the Law of Torts, traces the roots of such actions back to the early Roman law where the master sustained damages by the loss of a servant. Later, in 1349, after the Black Death left England with a severe labor shortage, a statutory remedy was enacted to meet the resulting agriculture crisis. An early English case, *Lumley v. Gye*, involved conduct to induce an opera singer of some distinction to breach her contract to sing exclusively at the plaintiff's theater. Although no fraud or violence occurred, the court sustained a right of action for inducing breach of contract but based its decision on a concept of malice that equates to an intentional tort.

Conduct within the area of business torts also includes improperly obtaining trade secrets, wrongfully appropriating another's good will, and interfering with business contracts and relationships. The Texas Supreme Court in *Raymond v. Yarrington*, 72 S.W. 580 (Tex. 1903), reasoned,

> It seems to us that where [plaintiff] has entered into a contract with another to do or not to do a particular act or acts, [plaintiff] has as clear a right to its performance as he has to his property, either real or personal; and that [defendant's action] knowingly to induce the other party to violate it is as distinctly wrong as it is to injure or destroy [plaintiff's] property.

Virtually any type of contract is sufficient as the foundation for an action for procuring its breach provided that the contract is in force and effect and not illegal or otherwise opposed to public policy so that the law would not aid in upholding it. Although most cases falling in the category of intentional interference customarily allege "malice," there is no satisfactory reason that would justify the courts reluctance to permit a remedy where the interference results from unintended negligence. However, in some cases the courts speak in terms of "proximate cause," indicating that the consequences are too "remote." The widely publicized alleged intentional invasion by Texaco in the 1980s, with the contractual rights of Pennzoil to acquire the assets of Getty, illustrates interference with contract rights in which the trial court awarded punitive damages of approximately $10 billion in addition to actual damages.

7.3 EVIDENCE OR PRESUMPTION OF NEGLIGENCE

Burden of Proof and Presumptions

Since the U.S. adversary system of litigation requires that evidence be produced by plaintiff and defendant, it is necessary that some methodology dispose of those cases in which the evidence is so inadequate or so conflicting that neither party can satisfy the trier of facts as to the truth of either version of the case. In short, in judicial resolution of disputes, one party must lose, and this "risk of non-persuasion" is referred to as the burden of proof. In civil cases, as opposed to criminal cases, the burden of proof does not require that the trier of facts (jury or court) be convinced beyond a reasonable doubt but generally by only a preponderance of the evidence in favor of the party with the burden of proof. In a negligence case, the burden of proving the defendant's negligence is on the plaintiff, since the plaintiff is asking for relief and must lose if the plaintiff's evidence does not meet the burden of proof by a preponderance of the evidence.

In certain cases a party is aided by procedural devices known as presumptions. A presumption is defined as "an assumption of the existence of a fact (fact A), which the law requires the trier of fact to infer as established, based on the existence of another fact or group of facts (fact B)." For example, as a general rule fact A ("the fact that X is dead") is presumed based on establishment of fact B ("that X has been unaccountably absent for a period of more than 7 years").

In the ordinary case, the plaintiff will produce evidence of the defendant's negligence by showing that the defendant should have known that his or her conduct reasonably would be expected to result in harm to the plaintiff. In *O'Hare v. Western Seven Trees Corp.*, 142 Cal. Rptr. 487 (1977), a landlord was held liable for injuries to the tenant in renting premises as "safe" when the landlord knew that there was a rapist in the area and said nothing to a prospective tenant.

In some negligence cases, plaintiffs may be able to take advantage of two doctrines to aid in meeting the burden of proof of negligence: negligence per se and *res ipsa loquitur.*

Violation of Statutes, Ordinances, and Regulations

The California Evidence Code Section 669(a) provides that a person is presumed to have failed to exercise due care if (1) the defendant violated a statute, ordinance, or regulation of a public entity; (2) the violation was the proximate cause of the plaintiff's injury; and (3) the injury resulted from an occurrence the nature of which the statute, ordinance, or regulation was designed to prevent.

The policy of this doctrine is that the standard of conduct required of a reasonable person may be properly established by legislative enactment. The statute providing that particular conduct is either prohibited or required may be interpreted as fixing a standard for the members of the community. A deviation may provide presumptive evidence of negligence, but, of course, proof of proximate cause also is necessary.

Some statutes are construed as creating no duty of conduct toward the plaintiff or duty with respect to a particular accident. For example, the Federal Safety Appliance Act requiring that railroads equip railcars with automatic couplers was judicially interpreted to cover injuries occurring in the course of coupling operations and not the risk of other casualties such as falls from the trains. By contrast, plaintiff pedestrian's injury when run down by defendant's automobile, driven through a stop sign at an excessive speed in violation of the speed limit, was presumed to be evidence of negligence by defendant. *Vallas v. Chula Vista*, 128 Cal. Rptr. 469, 472 (1976).

Res Ipsa Loquitur

In some cases, negligence may be difficult to prove because plaintiff lacks intimate knowledge of defendant's conduct or facilities. The expression *res ipsa loquitur* ("the thing speaks for itself") was introduced into Anglo-Saxon law by the 1863 case of *Byrne v. Boadle*, 159 Eng. Rep. 299 (Ex. 1863). In the *Byrne case*, a barrel of flour fell from the window of the defendant's warehouse and struck a passing pedestrian, the plaintiff. Although the plaintiff was unable to prove how or why the barrel rolled out of the window, the defendant was held liable upon a showing that the defendant was in possession of the warehouse. While the doctrine is not applicable to every unexplained accident, the justification for allowing the plaintiff to use *res ipsa loquitur* is that the affirmative evidence of the defendant's negligence is more than likely within the defendant's exclusive knowledge and control. In such cases, the defendant probably has superior knowledge of the facts leading up to the injury because of defendant's exclusive control of the object or facility that caused the harm. If the defendant was in fact negligent, defendant understandably will be reluctant to disclose facts that prove liability. However, where the defendant had exclusive control and the injury that occurred would not ordinarily happen in the absence of negligence, the doctrine creates an inference of negligence. A frequently quoted statement of the rule by the U.S. Supreme Court in *Sweeney v. Erving*, 33 S. Ct. 416 (1931), reads:

> *res ipsa loquitur* means that the facts of the occurrence warrant the inference of negligence, not that they compel such an inference; that they furnish circumstantial evidence of negligence where direct evidence of it may be lacking, but it is evidence to be weighed, not necessarily to be accepted as sufficient; that they call for explanation or rebuttal, not necessarily that they require it; that they make a case to be decided by the jury, not that they forestall the verdict.

Some courts go beyond the permissible inference theory and interpret the *res ipsa loquitur* doctrine as shifting the burden of presenting evidence by requiring the defendant to come forward and introduce evidence to rebut the inference of negligence. *Res ipsa* has been used frequently in instances of airplane crashes, falling elevators, or boiler explosions where the cause of the accident may be difficult for plaintiff to prove because important evidence was destroyed in the accident itself.

7.4 VICARIOUS LIABILITY

In searching for the policy considerations supporting the doctrine of vicarious liability (responsibility for the acts of another), several premises (i.e., human activity is destructive, automobiles kill more people than wars, excavations and building construction take a toll of passers-by, and erroneous credit reports destroy reputations) reflect useful benchmarks. In the early English case *Jones v. Hart*, 90 Eng.Rep. 1255 (1698), Chief Justice Holt starting with a proposition of law held that a person whose negligence caused harm should repair the harm, reasoned that "most of the people who operate machines which do the harm are too poor to pay for the losses they cause." Although the doctrine of vicarious liability has been severely criticized, there has been no split of authority or shifting back and forth since Holt's original statement of the doctrine. Subsequently, in discussing risk distribution, further support is supplied by the suggestions that "activities should bear the costs they engender and it is fair that an industry should pay for the injuries it causes." Hence, the notion of "enterprise liability" or "spreading the risk" was accepted rather than a principle of distribution solely on the basis of fault.

   Under the doctrine of vicarious liability, the employer is held responsible for torts committed by employees against a third party. Assuming no fault on the part of the employer or any wrongful intent, this resembles a form of strict liability as to the employer. While the employee is also responsible for the damages caused by the negligence, the plaintiff naturally prefers to sue the employer because of the employer's greater financial capability. The main thesis of the doctrine is not altered by permitting defenses against the injured plaintiff, such as contributory negligence and assumption of risk to be available to both the employee and the employer.

   The concept of vicarious liability is established firmly in the law of English-speaking countries. The criteria under which the doctrine has continued is based on (1) the principal/employer must have assented in some way to the relationship with the agent/employee, (2) the principal/employer must expect some benefit from the relationship, and (3) the principal/employer must exercise some element of "control." As the doctrine has been refined, its primary application is to fix responsibility on the employer business entity for the acts of the employee based on the policy that any person who is permitted to employ another person should bear corresponding responsibility including the "deep pocket" obligation to pay. The legal term "respondeat superior" means little more than "the buck stops here." It was reasoned that placing this responsibility on the employer permits the employer to distribute the cost including the expense of liability insurance through higher prices or rates to the public.

In fixing vicarious liability on an employer for the acts of another, an essential threshold determination is to establish the relationship of employer/employee as distinguished from that of an independent contractor relationship. Once that determination is made, the employer will be responsible for tortious conduct within the "scope of the employment." This nebulous formulation, sometimes also stated as "in the course of the employment," is legal shorthand to cover a range of both commanded and uncommanded acts that the employee performs for the employer. In general, the employee's conduct is within the scope of employment if the action is of the kind which the individual was employed to perform, occurs substantially within the authorized limits of time and space, and is actuated, at least in part, by a purpose to serve the employer.

In 1843, Baron Parke in an English case suggested a classic phrase that an employer was not liable for the torts of an employee who is not at all on the employer's business but is "going on the employee's own frolic." This phrase is now commonly referred to as the defense of "frolic and detour." Frequently, difficult questions of fact arise as to whether the employee's activity was an entire departure from the employer's business or merely carrying on the employer's business in a roundabout way. For example, this question is illustrated when the injury occurs while an employee deviates from the most direct route to engage in a personal errand or when an injury occurs while the employee is engaged in prohibited action or horseplay.

7.5 DEFENSES

General

The two traditional defenses that may be available to a negligent defendant are plaintiff's (1) contributory negligence or (2) assumption of risk. An underlying premise of the law is that all members of society have a duty to exercise reasonable care for their own safety. Accordingly, persons who fail to exercise reasonable care for their own safety are barred from recovery if their own negligence is a substantial contributing factor in producing their injuries.

Contributory Negligence

Contributory negligence is that conduct on the part of the plaintiff that (1) falls below the standard to which plaintiff should conform for plaintiff's own protection and (2) is a legally contributing cause, cooperating with the negligence of the defendant, resulting in the plaintiff's injury. For the purposes of determining contributory negligence, the standard of care for plaintiff is the same as that used in determining the negligence of the defendant (i.e., the amount of care that would be exercised by a reasonable person of ordinary prudence in similar circumstances). Likewise, the standard of care for a child or persons with reduced mental capacity is determined on the basis of the age, intelligence, and experience.

Since the doctrine of contributory negligence bars plaintiff's recovery, the results of applying the rule can be extremely harsh in some cases. Slightly negligent persons may be unable to recover anything, although they have received very serious injuries as the result of the defendant's negligence. As a result, some states adopted systems of comparative negligence, which is discussed in a later paragraph of this section. Since the system of comparative negligence is of statutory or judicial creation, the details of the various systems differ among the states in their details. However, the basic idea of comparative negligence is to diminish the plaintiff's recovery according to the degree of plaintiff's fault rather than a complete bar to recovery.

A number of courts also hold that the contributory negligence of the plaintiff will not bar recovery if the plaintiff can show that the defendant had the "last clear chance" to avoid the harm. The idea of last clear chance is that, despite the contributory negligence of the plaintiff, if the accident could have been avoided by the subsequent exercise of reasonable care by the defendant, that superior opportunity to avoid the accident makes the defendant's last clear chance the critical fault.

Assumption of Risk

Under the doctrine of assumption of risk, the plaintiff is prohibited from recovering even though the defendant has been negligent. Such advance consent by the plaintiff to assume the risk relieves the defendant of any obligations toward the consenting party. The plaintiff's consent may be either express or implied. The issues raised in assumption of the risk defenses usually turn on questions of whether the parties were of substantially equal bargaining power and whether the plaintiff knowingly and freely consented to the risk.

A California case, *Fonseca v. County of Orange*, 104 Cal. Rptr. 566, 104 P.2d 566, (Cal.App. 1972), involved a plaintiff construction worker who slipped on wet cement and fell from a bridge under construction when the plaintiff was doing finishing work on the deck of the bridge some 20 feet above a dry river bed. The contractor had failed to install a scaffolding as required by law when workers were working at heights in excess of 7.5 feet. Although the employer sought to defend on the grounds of contributory negligence and assumption of risk, both defenses were rejected. The court determined that contributory negligence was not a defense to injuries caused by violations of safety regulations. In dismissing the defense of assumption of risk, the court's decision was grounded on the employee's weak bargaining position resulting from economic pressure upon the employee to work in an unsafe place or with an unsafe appliance.

Comparative Negligence

In 1975, an important decision of the California Supreme Court, representing a general trend throughout the United States, abolished the rule of contributory negligence that existed in the law of California for a century. In *Li v. Yellow Cab Co.*, 13 Cal.3rd 804, 828, 119 Cal. Rptr. 858, 875 (Cal. 1975), the court rejected the doctrine of "all or nothing," and held that the damages awarded to the plaintiff should not be barred but should be diminished in proportion to the amount of negligence attributable to the person recovering. At the same time, California abolished the doctrine of last clear chance, which created an exception to the rule that contributory negligence barred recovery.

The rule of comparative negligence in effect seeks to view the event of the injury in totality and then answer the question of whether or not it is fair to allow the plaintiff full compensation for an injury in which the plaintiff's negligence was a significant factor. By adopting this rule, the court squarely makes a comparison of fault relevant by looking at the conduct of both the plaintiff and defendant. The draft Uniform Comparative Fault Act proposed by the National Conference of Commissioners on Uniform State Laws provides

> (a) In an action based on fault, to recover damages for injury or death to person or harm to property, any contributory fault chargeable to the claimant diminishes proportionately the award of compensatory damages, but does not bar recovery . . . .

> (b) Fault includes negligence, recklessness, breach of implied warranty, conduct subjecting the actor to strict tort liability, unreasonable assumption of risk, and failure to avoid or mitigate damage. The fault must have an adequate causal relationship to the damage suffered.

Adoption of the comparative negligence system in California also entails a merger of the assumption of risk doctrine in the system of comparative negligence by treating the plaintiff's conduct as negligence in proportion to the fault. *Gonzalez v. Garcia*, 75 Cal. App.3rd 874, 877, 142 Cal. Rptr. 503, 505 (Cal. 1977).

7.6 STATUTE OF LIMITATIONS

The so-called statute of limitations deals with the effect of passage of time upon the ability of the claimant to maintain a legal action to recover for the alleged injury. These statutes, sometimes referred to as a defense, apply to all lawsuits whether contract or tort and are designed to protect defendants from false or fraudulent claims that may be difficult to disprove if not brought in a timely manner. The policy back of the statutory rule is that the relevant evidence may be lost or destroyed and witnesses may be unavailable or the events may have faded from their memories. In addition, statutes of limitations have another purpose since the law seeks to promote certainty and finality to transactions and relationships between persons and business entities.

Despite the affirmative desirability of these statutes, their implementation has caused many problems in certain areas, especially negligence actions involving construction projects and certain personal injuries. In construction projects, claims are frequently discovered long after the construction is completed, and the statutes are not clear as to when the statutory period barring the claim begins to run. Although the statutory language is cast as a period "within 4 years from the time the cause of action accrued," this formulation begs the question. For example, in a construction project, it is not clear whether the cause of action accrued (1) when the wrongful act (negligent design or construction) occurred, (2) when the injury in fact occurred, or (3) when the claimant knew or should have known of the claim (defect or negligence). In an effort to clarify the situation, a number of states adopted "completion statutes" seeking to cut off liability in a specific number of years after "substantial completion." However, these statutes vary in their details with respect to the persons who could take advantage of the statute as well as differentiating between latent and patent defects. Some states have not looked favorably on such statutes that sought to cut off persons rights when they did not know a claim existed. *Shibuya v Architects Hawaii Limited*, 647 P.2d 276 (Hawaii 1982). This difficulty may be avoided by contractual waivers such as are contained in AIA Doc. B141:

. . . as to all acts or failures to act by either party to this agreement, any applicable statute of limitations shall commence to run and any alleged cause of action shall be deemed to have accrued in any and all events not later than the relevant Date of Substantial Completion of the Work . . . .

The numerous cases involving exposure to asbestos illustrate some practical difficulties arising in personal injury cases. Generally, the statute of limitations for such personal injury cases seeks to resolve the question by requiring that suit must be brought within 3 years after injury. However, the asbestosis cases demonstrate the failure to solve the question since several possible interpretations include (1) initial exposure, (2) date of last exposure (under which the claim could be barred before the injury becomes manifest), or (3) date of manifestation of injury (i.e., either first impairment or the date the first x-ray showed an abnormality). This date of manifestation of injury variation relates to the so-called "discovery rule," which provides that the statute begins to run from the date the victim knew or should have reasonably known of the injury from a particular cause or source. *DeMato v. Turner & Newall, Ltd.* 651 F.2d 908, (3rd Cir. (1981).

7.7 DAMAGES

The principal function of awarding tort damages is to compensate the injured plaintiff for the loss caused by injury. Ordinarily, the plaintiff is entitled to recover economic losses and certain noneconomic losses.

Damages for economic losses may include in cases of personal injuries, the amount of lost earnings and medical expenses; in cases involving the diminution in value of damaged property, the recovery may include the cost of repair or replacement of the property; and in cases of fraud or misrepresentation, the damages may reflect losses attributable to the benefit of bargain or out- of-pocket expense. With respect to noneconomic loss, recovery for emotional distress is granted hesitantly where the principal element is pain and suffering.

In a sense, the underlying ideas of criminal law have invaded the field of torts where the defendant's conduct was intentional and deliberate reflecting the character of an outrage. In such cases, the courts have awarded the plaintiff "punitive" or "exemplary" damages. The rationale for awarding such damages in addition to full compensation for the plaintiff's injuries is explained on the grounds of teaching the defendant a lesson and deterring others from following the defendant's example.

There are two major limitations on the recovery of tort damages including the avoidable consequences rule and those caused by the plaintiff's fault. Plaintiff's failure to mitigate or minimize the damages caused by defendant's negligence is a defense, treating such conduct as negligence bearing on the proportionate reduction in damages. *LeMons v Regents of University of California*, 148 Cal. Rptr. 335, 582 P.2d 946 (Cal. App. 1978). As was discussed earlier, in most jurisdictions, a plaintiff may recover full compensation for all losses under the comparative negligence rule only if the plaintiff is found free of contributory negligence.

7.8 ENGINEERING ETHICS AND NEGLIGENT CONDUCT

The term "technethics" is defined in *New Words and Meanings* of the *Britannica Book of the Year* (Encyclopedia Britannica, Inc., Chicago, Ill., 1973) as the responsible use of science, technology, and ethics in a society shaped by technology. The definition seems to suggest a prioritization into a dominant and subordinate factor by describing "society shaped by technology" while involving "responsible use of ethics." The thrust of this seeming skew of policy objectives, the ethical aspects of technological advances, receive only secondary recognition. Ethical considerations must emerge from only passive "gentlemanly" conduct to real concern for public welfare. Thus, engineers must assess the impact of "unsafe at any speed," "calculated health risk," and "acceptable pollution" in terms of their ultimate public damage impact on society. While the need for a high degree of specialization in the training of engineers is generally accepted, those responsible for engineering curriculum should include sufficient cross-disciplinary courses to assist engineering students in evaluating the relationship of their technical and scientific achievements in relation to societal questions. In the more noble calling of professionalism, the engineer must look beyond the day-to-day performance of the technical or scientific aspect of work to realize the broader consequences of technological advances measured by the implications of that progress in terms of ultimate human values.

A positive case can be made out for diligent self-discipline through the adoption and enforcement of professional codes of ethics. A "crabbed interpretation" and casual respect for ethical conduct can only hasten popular demand for statutory reform and increased liability. See, *Calvert Cliffs Coordinating Committee v. United States Atomic Energy Commission*, 449 F.2d 1109 (D.C. Cir. 1971). Failure to respond to societal goals through self-discipline will necessarily result in the progressive movement of legal doctrine in its more rigid form. Just as legal standards of acceptable conduct in answer to societal changes moved from intentional torts to negligent torts to strict liability, so too societal pressure as the arbitrator will provide mandatory reform to achieve developing human values.

ARTHUR v. STANDARD ENGINEERING CO.
193 F.2d 903 (D.C. 1951)

Wilbur K. Miller, Circuit Judge. [Plaintiff], an employee of an electrical subcontractor, sues for injuries sustained in a fall from scaffolding which had been installed by employees of a steam fitting subcontractor to do their steamfitting work. Plaintiff and other employees asked and were given permission to use the scaffolding by the defendant company. Plaintiff fell about 15 feet from the scaffolding when a long supporting plank broke. Plaintiff sustained serious injuries and brought an action to recover from the steamfitting subcontractor.

It is well settled that an invitor owes his invitee the duty of furnishing him with reasonably safe premises or appliances and will be held liable for injuries caused by a negligent failure to perform the duty. It is equally well established that a licensor is not liable in damages for the injuries of a mere licensee, unless the injuries were the result of the licensor's active negligence . . . .

The case turns, therefore, on the question whether an employee of a subcontractor is in the legal status of a licensee or that of an invitee when he uses, either by permission expressly given or implied from established custom a scaffold erected by the employees of another subcontractor for their own use.

The Supreme Court in *Bennett v. Louisville L & N R.R. Co.* [stated]

> The principle . . . appears to be that invitation is inferred where there is a common interest or mutual advantage, while a license is inferred where the object is the mere pleasure or benefit of the person using it . . . .

We adopt, therefore, the mutual benefit test for determining whether the plaintiff was a licensee or an invitee . . . . The true test under the mutual advantage theory is whether the owner of the scaffold or other appliance receives benefit or advantage from the permitted use by another of that particular piece of equipment. If so, the user is an invitee; if not, he is a licensee . . . . We conclude that plaintiff was a licensee. For Defendant.

KENTUCKY & WEST VIRGINIA POWER CO. v. ANDERSON
156 S.W.2d 857 (Ky. 1941)

Stanley, Commissioner. [Plaintiff], Anderson, who lives adjacent to the electric substation of the defendant, Kentucky & West Virginia Power, brings this action to recover damages for a nuisance. Plaintiff acquired a five room frame house 19 years before the defendant purchased the adjacent property for use as a substation and installed four transformers about 7 feet from plaintiff's fence and 15 feet from plaintiff's house.

The transformers consist of a large iron core surrounded by a series of wires which reduce the current by induction caused by the magnetizing and demagnetizing 120 times a second. This causes a vibration and humming. Persons in plaintiff's house are disturbed in their conversations and in their sleep by this constant buzzing sound. In addition the use of a radio is practically destroyed. The court instructed the jury that plaintiff was entitled to recover if "the noises constantly and continuously emanating from the transformers [were of] such character and degree as to produce actual physical discomfort and annoyance to a person of ordinary health and average sensibilities . . . ."

There can be no doubt but that commercial and industrial activity which are lawful in themselves may become nuisances if they are so offensive to the senses that they render the enjoyment of life and property uncomfortable. It is no defense that skill and care have been exercised and that the most improved methods and appliances employed to prevent such result. Of course, the creation of trifling annoyances and inconvenience does not constitute an actionable nuisance, and the locality and surroundings are of importance. The determining factor when noise alone is the cause of complaint is not the intensity or volume. It is that the noise is of such character as to produce actual physical discomfort and annoyance to a person of ordinary sensibilities, rendering adjacent property less comfortable and valuable. If the noise does that, it can well be said to be substantial and unreasonable in degree; and reasonableness is a question of fact dependent upon all the circumstances and conditions. There can be no fixed standard as to what kind of noise constitutes a nuisance. It is true some witnesses in this case say they have not been annoyed by the humming of these transformers, but that fact is not conclusive as to the nonexistence of the cause of complaint, the test being the effect which is had upon a ordinary person who is neither sensitive nor immune to the annoyance concerning which the complaint is made. In the absence of evidence that the complainant and his family are super sensitive to distracting noises, it is to be assumed that they are persons of ordinary and normal sensibilities. For Plaintiff.

VOSS v. KINGDON & NAVEN, INC.
328 N.E.2d 297 (Ill. 1975)

[Scaffold Acts have been adopted in several States. These statutes impose on parties deemed *in charge* of a construction project a duty to exercise care for the protection of persons on the project. The Scaffold Acts do not impose strict liability; negligence must be proved. Plaintiff was injured when scaffolding collapsed at a sewage disposal plant project. Under contract among the City, contractor (plaintiff's employer) and the defendant engineering firm, the defendant's engineer was to provide inspection to assure compliance with the contract documents. The defendant engineering company employed a resident engineer who had power to stop the work. The evidence indicated that the engineer knew one of the vertical supports for the scaffolding had been removed earlier in the day. The trial court directed a verdict for the defendant engineering firm.]

The contract between the defendant and the city provided:

8-8 Suspension of Work

The engineer shall have authority to suspend the work, wholly or in part, for each period of time as he may deem necessary, due to conditions unfavorable for the satisfactory prosecution of the work, or if conditions in his opinion warrant such action; or for such time as is necessary by reason of failure on the part of the Contractor to carry out orders given, or to perform any or all provisions of the contract.

Sect. 9 Structural Work Act

Any owner, contractor, subcontractor, foreman or other person having charge of the erection, construction, repairing, alteration, removal or painting of any building, bridge, viaduct or other structure within the provisions of this Act, shall comply with all of the terms thereof.

For any injury to person or property occasioned by willful violations of any of [the provisions of the Structural Work Act], a right of action shall accrue to the party injured for any direct damages sustained.

The court earlier addressed itself as to what is meant by "having charge of" under the Structural Work Act. It was said that the term "having charge of" is a generic term of broad import, and although it may include supervision and control, it is not confined to it. Consistent with its beneficial purpose of preventing injury to persons employed in the extra-hazardous occupation of structural work, the thrust of the statute is not confined to those who perform, or supervise, or control, or retain the right to supervise or control, the actual work from which the injury arises, but, to insure maximum protection, is made to extend to owners and others who have charge of the erection or alteration of any building or structure.

Under the contract among the City, the contractor and the defendant (which contract had been drawn by defendant), the contractor would be obliged at the defendant's request to discharge or remove workmen whose work the defendant considered to be improper. Too, defendant was given authority to suspend work wholly or in part for such time as it might consider necessary because of "conditions unfavorable for the satisfactory prosecution of the work," or "due to conditions which in his opinion" warranted such suspension of work. And the defendant was given authority to suspend the work if there was any failure of the contractor to carry out orders given or to perform any provision of the contract.

Judgment of the trial court reversed. New trial order.

IRA S. BUSHEY & SONS, INC. v. UNITED STATES
398 F.2d 167 (2nd Cir. 1968)

Friendly, Circuit Judge. While the United States Coast Guard vessel Tamaroa was being overhauled in a floating drydock located in Brooklyn's Gowanus Canal, a seaman returning from shore leave late at night, in the condition for which seamen are famed, turned some wheels on the drydock wall. He thus opened valves that controlled the flooding of the tanks on one side of the drydock. Parts of the drydock sank, and the ship partially did—fortunately without loss of life or personal injury. The drydock owner was granted compensation and the United States appeals.

The Government attacks imposition of liability on the grounds that Lane's acts were not within the scope of his employment. It relied heavily on Section 228 of the RESTATEMENT of Agency 2d which says that "conduct of a servant is with the scope of employment if, but only if: . . . (c) it is actuated, at least in part by a purpose to serve the master."

It would be going too far to find such a purpose here; while Lane's return to the Tamaroa was to serve his employer, no one has suggested how he could have thought turning the wheels to be, even if—which is by no means clear—he was unaware of the consequences.

In the light of the highly artificial way in which the motive test has been applied, the district judge believed himself obliged to test the doctrine's continued vitality by referring to the larger purposes *respondeat superior* is supposed to serve. He concluded that the old formulation failed this test. We do not find his analysis so compelling, however, as to constitute a sufficient basis in itself for discarding the old doctrine. It is not at all clear, as the court below suggested that the expansion of liability in the manner here suggested will lead to a more efficient allocation of resources. As the most astute exponent of this theory has emphasized, a more efficient allocation cannot be expected if there is some reason to believe that imposing a particular cost on the enterprise will lead it to consider whether steps should be taken to prevent an occurrence of the accident. Calabresi, "The Decision for Accidents: An Approach to Non-fault Allocation of Costs," 78 *Harv.L.Rev.* 713, 725-34 (1965). And the suggestion that imposition of liability here will lead to more intensive screening of employees rests on highly questionable premises. The unsatisfactory quality of the allocation of resources rationale is especially striking on the facts of this case. It could well be that application of the traditional rule might induce drydock owners, prodded by their insurance companies, to install locks on their valves to avoid similar incidents in the future, while placing the burden on the shipowners is much less likely to lead to accident prevention. It is true, of course, that in many cases the plaintiff will not be in a position to insure, and so expansion of liability will, at the very least, serve *respondeat superior's* loss spreading function. But the fact that the defendant is better able to afford damages is not alone sufficient to justify legal responsibility and the overarching principle must be taken into account in deciding whether to expand the reach of *respondeat superior.*

A policy analysis thus is not sufficient to justify this proposed expansion of vicarious liability. This is not surprising since *respondeat superior,* even within its traditional limits, rests not so much on policy grounds consistent with the governing principles of tort law as in a deeply rooted sentiment that a business enterprise cannot justly disclaim responsibility for accidents which may fairly be said to be characteristic of its activities. It is in this light that the inadequacy of the motive test becomes apparent. Whatever may have been the case in the past, a doctrine that would create such drastically different consequences for the action of the . . . drunken seaman here reflects a wholly unrealistic attitude toward the risks characteristically attendant upon the operation of a ship.

Put another way, Lane's conduct was not so "unforeseeable" as to make it unfair to charge the Government with responsibility. We agree with a leading treatise that "what is reasonably foreseeable in the context of [*respondeat superior*] is quite a different thing from the foreseeably unreasonable risk of harm that spells negligence . . . . The foresight that should impel the prudent man to take precautions is not the same measure as that by which he should perceive the harm likely to flow from his long-run activity in spite of all reasonable precautions on his own part. The proper test here bears far more resemblance to that which limits liability for workmen's compensation than to the test for negligence . . . . 2 Harper & James, *The Law of Torts*, 1377-78 (1956).

Consequently, we can no longer accept our past decisions that have refused to move beyond [the so-called *Nelson rule*, announced in *Brailas v. Shepard S.S. Co.*, 152 F.2d 849 (2d Cir. 1945), cert. denied 66 S. Ct. 970, (1946). Since that rule does] not accord with modern understanding as to when it is fair for an enterprise to disclaim the actions of its employees.

One can readily think of cases that fall on the other side of the line. If Lane had set fire to the bar where he had been imbibing or had caused an accident on the street while returning to the drydock, the Government would not be liable; the activities of the "enterprise" do not reach into areas where the servant does not create risks different from those attendant on the activities of the community in general . . . . Affirmed.

DISCUSSION QUESTIONS

1. Define and distinguish (1) contractual liability, (2) criminal liability, and (3) tort liability.

2. Define and illustrate (1) negligence, (2) gross negligence, (3) contributory negligence, (4) assumption of risk, and (5) last clear chance.

3. What and who may be liable if an injury occurs at a construction site as a result of inadvertence such as failure to cover holes or leaving a ladder standing in a dangerous position?

4. How may an owner become responsible for injuries to employees of the general contractor who was considered to have the legal status of "an independent contractor"?

5. Discuss the economic and social policy aspects of "contributory negligence" and "comparative negligence."

6. In the case of a well-meaning but misguided samaritan, who upon seeing a stalled motorist applies an unsought bumper-push propelling the stalled car into the intersection causing a collision with another vehicle entering the intersection, may the owner of the other vehicle recover from the owner of the stalled vehicle? With respect to your answer discuss the elements of (1) assent, (2) benefit, and (3) control.

8

PRODUCT LIABILITY

8.1 BACKGROUND

As late as the nineteenth century, manufacturers were generally shielded from liability resulting from the production of defective or dangerous products owing to the reluctance of the courts to find a legal basis for liability. The courts justified their reluctance to establish liability on the premise that holding the manufacturer responsible would result in (1) grave administrative complications and (2) adverse social consequences causing economic ruin of product suppliers. The implication of this judicial approach placed the burden on consumers to ferret out inferior and defective products or correct product weaknesses and deficiencies after purchase.

A landmark case in this development was *Winterbottom v. Wright*, 152 Eng. Rep. 402 (Ex. 1842), which involved a suit by plaintiff, a driver of a mail coach, who was injured as the result of defendant's faulty maintenance of the coach. Plaintiff's employer, Atkinson, received the mail coach from the postmaster general. In turn, the defendant had a contract with the postmaster general under which defendant agreed "to keep the coach in a fit, safe, and secure condition." As the result of defendant's negligent maintenance, plaintiff driver was injured when the coach broke down and plaintiff was thrown to the ground. The court dismissed plaintiff's complaint based on two separate lines of reasoning: administrative and conceptual.

The thrust of the administrative objection was characterized by the court as opening a floodgate of litigation:

> If the plaintiff can sue, [then] every passenger, or even any person passing along the road, who was injured by the upsetting of the coach, might bring a similar action. Unless we confine the operation of such contracts as this to the parties who entered into them, the most absurd and outrageous consequences, to which I can see no limit, would ensue. . . . The only safe rule is to confine the right to recover to those who enter into the contract; if we go one step beyond that, there is no reason why we should not go fifty.

The argument of administrative complications proved groundless and was not supported by the feared floodgate of product liability litigation, but, rather, an extensive volume of litigation was spawned separately by the procedural acceptance of class action lawsuits which are not limited to the field of product liability litigation. Class actions are lawsuits in which all similarly situated plaintiffs may join in a single lawsuit or in which representative plaintiffs may sue on behalf of all plaintiffs similarly situated. These class action lawsuits gained widespread popularity outside of the product liability field in a variety of other cases such as securities fraud, stockholder complaints, and consumer credit.

Since, apparently, the postmaster general did not sue the defendant, the conceptual argument was articulated that permitting plaintiffs to sue without privity of contract was unsound as a matter of justice:

> [B]y permitting this action, we should be working this injustice, that after the defendant had done every thing to the satisfaction of his employer [the postmaster general], and after all matters between them had been adjusted, and all accounts settled on the footing of their contract, we would subject [the defendant] to be ripped open by this action of tort brought [by the plaintiff driver] against him. The duty, therefore, is shown to have arisen solely from the contract; and the fallacy consists in the use of the word "duty." If a duty to the postmaster general be meant, that is true; but if a duty to the plaintiff be intended (and in that sense the word is evidently used), there was none.

This lack of privity of contract argument remained the primary line of defense urged by manufacturers based on the reasoning that in the absence of a contractual relationship between the injured party and the manufacturer no action could be maintained by the injured party against the manufacturer. That is, the ultimate consumer, who did not purchase the product directly from the manufacturer but purchased it down the distribution chain from a retailer, could not bring a legal action against the manufacturer for injuries caused by defective products. However, courts have looked with disfavor on this defense in product liability cases, and, since the turn of the century, the courts have whittled away at the doctrine of privity so that under the recent trend of cases, the limitation has almost completely eroded. Representative of the trend seeking to avoid the "privity limitation" is *Wisdom v. Morris Hardware Co.*, 274 P. 1050 (Wash. 1929). In that case, when a plaintiff sought damages from the manufacturer for injury to a fruit crop because of the use of a certain spray purchased through a retailer, the court rejected the defendant's lack of privity defense, finding an agency relationship existed between the manufacturer and the retailer by reasoning that the manufacturer, as principal, was responsible since the retailer was acting as defendant manufacturer's agent.

Judicial impediments to the plaintiff's recovery were gradually chipped away, by bringing lawsuits founded on contractual warranties (express and implied), negligence and strict liability, and by relating liability to (1) inherently dangerous products and (2) the defendant's knowledge of or efforts to conceal the defects.

Demise of the Privity Requirement

In an early twentieth century New York case, *McPherson v. Buick*, 111 N.E. 1050 (N.Y. 1916), involving an automobile and not an inherently dangerous product such as poison or explosives, Justice Benjamin Cardozo speaking for the court rejected the privity requirements. In that case, the plaintiff, while driving a car purchased from a retail distributor and manufactured by defendant car company (Buick), was injured when one of the wheels proved defective. The wheel was not manufactured by defendant but was one of some 80,000 wheels which defendant purchased from a reputable contractor, none of which had proved to be made of defective wood prior to the accident in the present case. Justice Cardozo, speaking for the court,

> If the nature of a thing is such that it is reasonably certain to place life and limb in peril when negligently made, it is then a thing of danger. Its nature gives warning of the consequences to be expected. If to the element of danger there is added knowledge that the thing will be used by persons [ultimate consumer] other than the purchaser [retail dealer] and used without new tests, then, irrespective of contract, the manufacturer of this thing of danger is under a duty to make it carefully.

Dissenting from the majority holding for the plaintiff, Justice Barlett relying on the *Winterbottom case* noted,

In the case at bar the defective wood on an automobile moving only eight miles an hour was not any more dangerous to the occupant of the car than a similarly defective wheel would be to the occupants of a carriage drawn by a horse at the same speed; and yet unless the courts have been all wrong on this question up to the present time there would be no liability to strangers to the original sale . . . .

Inherently Dangerous Products

Concern about the breadth of the manufacturers' liability was expressed in a case where the plaintiff was injured as a result of falling through a metal threshing machine cover which was insufficient to hold the weight of a man although it was necessary to stand on the cover to operate the machine. In commenting on the issue of remoteness of damage or foreseeability, the court stated,

[P]erhaps more than all this, [the action is never allowed] for the reason that a wise and conservative public policy has impressed the courts with the view that there must be a fixed and definite limitation to the liability of manufacturers and vendors for negligence in the construction and sale of complicated machines and structures which are to be operated or used by the intelligent and the ignorant, the skillful and the incompetent, the watchful and the careless; parties that cannot be known to the manufacturers or vendors, and who use the article all over the country hundreds of miles distant from the place of their manufacture or original sale . . . .

While the "privity doctrine" was cast in terms of contractual relationships, proper analysis should cast the issue as one of negligence and in terms of either breadth or limitation of a "duty." Under the *Winterbottom case*, it was held that no duty could run to the plaintiff on the strength of a contractual undertaking to a third party, in that case the defendant's immediate vendee/postmaster general. Cardozo opened the judicial door to cast product liability litigation in terms of "use by the public" of products that are "inherently dangerous."

The early case of *Thomas v. Winchester*, 6 N.Y. 397, (1852) involved a drug that was easily categorized as an inherently dangerous product. There, the plaintiff (ultimate consumer) was injured by taking a small quantity of belladonna that had been mislabeled by an employee of the defendant (wholesaler) as dandelion extract. The former is a "poison," the latter is a "harmless medicine." The plaintiff's case turned on the fact that plaintiff purchased the belladonna from a local druggist who in turn had purchased it from the defendant wholesaler. On the basis of the fact that the dangerous drug remained in its original condition until it caused the plaintiff's injury, coupled with the absence of any culpability in the distribution chain or in the plaintiff's use, the court transformed the analysis to a causation test that the defendant is responsible for "the natural and necessary" consequences of its negligence. The opinion noted,

> In the present case the sale of the poisonous article was made to a dealer in drugs, and not to a consumer. The injury therefore was not likely to fall on him (wholesaler), or on his vendee who was also a dealer; but much more likely to be visited on a remote purchaser, as actually happened.

In an unusual decision, the California supreme court in *Sindell v. Abbott Laboratories*, 607 P.2d 924 (Cal. 1980), held the entire drug manufacturing industry responsible for plaintiff's cancer allegedly caused by plaintiff's mother taking a prescribed medicine diethylstilbestrol (DES) to prevent a miscarriage. There were no records available (20 years after the drug was taken) to show which manufacturer supplied the drug, but five large manufacturers supplied the same formula and collaborated in marketing the drug. Describing this as "marketshare" liability, the court procedurally held that defendants had the burden of proof to show that it (any manufacturing company) could not have been the supplying manufacturer.

Defendant's Knowledge of or Concealment of Defect

The trend of the law in negligent injury product liability cases resulted in questioning the "subjective" test as to the defendant's knowledge of the product defect and substituted the "objective" standard of reasonable prudent conduct of the manufacturer. This judicial change holds the defendant responsible for harm caused, even though the defendant did not have personal knowledge of the defect, as long as that harm could have been prevented by a reasonable person in defendant's situation.

In *Greenman v. Yuba Power Products, Inc.*, 27 Cal. Rptr. 697, 377 P.2d 897 (Cal. 1963), the plaintiff was injured while using a Shopsmith combination power tool (which was purchased by the plaintiff's wife) and manufactured and marketed by the defendant. While the tool was being used as a lathe, a piece of wood suddenly flew out of the machine and struck the plaintiff on the forehead. The complaint alleges that the Shopsmith had defective set screws with insufficient strength to hold the wood properly. The court held that the machine's presence on the market was tantamount to a "representation" supporting an implied warranty that "the machine would safely do the jobs for which it was built." Justice Traynor speaking for the court held,

In my opinion it should be recognized that a manufacturer incurs an absolute liability when an article that he places on the market, knowing that it is to be used without inspection, proves to have a defect that causes injury to human beings.

Recognized first in the case of unwholesome food products, such liability has now been extended to a variety of other products that create as great or greater hazards if defective.

This extension of the policy as to the duty owed the purchaser/user, in effect, equates to strict liability on the theory that even if there is no negligence, public policy demands that the responsibility be fixed wherever it will most effectively reduce the hazards to life and health inherent in defective products that reach the market. This judicial doctrine reasons that the manufacturer can anticipate some hazards and guard against the recurrence of others, as the public cannot. Further, in balancing the costs to the injured party and the manufacturer, some courts reason that the cost to the injured party may be an overwhelming misfortune, while the manufacturer may obtain protection by insurance that can be distributed among the public as a cost of doing business.

8.2 THEORIES OF PRODUCT LIABILITY CASES

Negligence

When an accident occurs in a situation implicating a product as a possible link in the cause of injury, the injured party begins to assess whether the law will hold liable either the manufacturer or the seller of the product. On a straight negligence theory, no legal doctrine imposes *ipso facto* liability on the grounds that a product had some direct or indirect relationship to the injury since the manufacturer or seller is not an insurer. Absent the use of strict liability, the product liability plaintiff seeking to recover on a negligence theory must prove that the manufacturer or seller of the product violated some legal duty of care owed to the injured party.

Negligence is defined as "conduct falling below the standard established by the law for the protection of others against unreasonable risk of harm." As was discussed in greater detail in chapter 7, the plaintiff is not required to prove that the defendant either intended harm or acted recklessly. It is important to note that recovery under the theory of negligence focuses on the *conduct* of the defendant (manufacturer or seller). The test relates to a norm of conduct referred to as a "reasonably prudent person."

Humans are constantly subjected to elements of risk but are only liable when their conduct falls below what a "reasonable person" would have done under similar circumstances. This analysis calls for an examination of whether the defendant (manufacturer or seller) failed to exercise "reasonable care" as demonstrated by quality control, warning labels, and other factors. In making that decision, consideration must be given to (1) the probability of injury occurring, (2) the nature or gravity of the harm that may result, and (3) the precautionary steps taken to avoid or minimize such injury.

In focusing on the defendant's conduct, it is relevant to examine not only the technology known but also the information available to the manufacturer concerning the risks inherent in the product. In *Marsh Wood Products Company v. Babcock and Wilcox*, 240 N.W. 392, (Wisc. 1932), plaintiff was injured by an exploding boiler that defendant had manufactured and tested by using a hydrostatic test to establish the boiler's integrity. Plaintiff recovered on the grounds that a more sophisticated test, recently developed by a professor in the metallurgical department at the University of Wisconsin, provided a higher probability of discovering flaws in the metal than the test used by the defendant. While the defendant may attempt to negate evidence of negligence by proof of compliance with industry or governmental standards, a line of cases holds that "reasonable care" may require more vigilant behavior and a greater degree of care than meeting governmental or industry guideline standards. *Berkebile v. Brantly Helicopter Corp.*, 281 A.2d 707 (Pa. 1971).

Strict Liability

The concept of strict liability emerged as a sound legal doctrine in 1964 with the approval in the RESTATEMENT (Second) of Torts, Section 402A, which reads,

*Special Liability of Seller of Product
for Physical Harm to User or Consumer*

(1) One who sells any product in a defective condition unreasonably dangerous to the user or consumer or to his property is subject to liability for physical harm thereby caused to the ultimate user or consumer, or to his property, if

    (a) the seller is engaged in the business of selling such a product, and

(b) it is expected to and does reach the user or consumer without substantial change in the condition in which it is sold.

(2) The rule stated in subsection (1) applies although

(a) the seller has exercised all possible care in the preparation and sale of his product, and
(b) the user or consumer has not bought the product from or entered into a contractual relation with the seller.

CAVEAT:

The Institute expresses no opinion as to whether the rules in this Section may not apply

(1) to harm to persons other than users and consumers;

(2) to the seller of a product expected to be processed or otherwise substantially changed before it reaches the user or consumer; or

(3) to the seller of a component part of a product to be assembled.

Under the analysis of strict liability, the focus is on the product as contrasted with the defendant's conduct in the case of negligence. In the leading case of *Escola v. Coca Cola Bottling Company*, 24 Cal.2d 497, 150 P.2d 436, (Cal. 1944), plaintiff waitress in a restaurant was injured by an exploding bottle of Coca Cola when it was removed from the refrigerator. Plaintiff alleged "that the bottles contained excessive pressure of gas or by reason of a defect in the bottle was dangerous and likely to explode." Evidence was introduced to show that the bottles were delivered to the restaurant some 36 hours earlier and were placed in the refrigerator. Plaintiff did not offer any evidence as to the defendant's negligence in the manufacture and preparation of the bottle. The defendant through an expert witness testified as to the procedures under which the bottles were made and marketed. In holding for the plaintiff, the court reasoned squarely on the doctrine of *res ipsa loquitur* supported by "evidence permitting a reasonable inference that [the bottle] was not accessible to extraneous harmful forces and that it was handled carefully by plaintiff or any third party who may have moved or touched it."

In distinguishing the theory of strict liability from the doctrine of negligence, in the former it is immaterial that the manufacturer met the standard of reasonableness by utilizing all possible care in quality control techniques. If the product is "unreasonably dangerous" and it caused the plaintiff's injury, the manufacturer is liable.

Breach of Warranty and Misrepresentation

The manufacturer or seller of a product may be liable to a person injured in the use of that product under the theory of a breach of express or implied warranty. Section 2-313 of the Uniform Commercial Code (UCC), adopted in all states except Louisiana, provides

(1) Unless excluded or modified, a warranty that the goods shall be merchantable is implied in a contract for their sale if a seller is a merchant with respect to goods of that kind . . . .

(2) Goods to be merchantable must be at least such as . . . .

(c) are fit for the ordinary purposes for which such goods are used . . . .

As a condition of plaintiff's recovery under the UCC warranty theory, the UCC requires notice by the injured party to the manufacturer or seller, and failure to give timely notice may bar recovery. Under a warranty theory, the focus is on a contractual breach rather than on some form of product defect. To recover damages, the plaintiff must establish that injury resulted from the failure of the product to meet the warranty.

Section 2-313 of the UCC dealing with express warranties, states

(1) Express warranties by the seller are created as follows:

(a) Any affirmation of fact or promise made by the seller to the buyer which related to the goods and becomes part of the basis of the bargain creates an express warranty that the goods shall conform to the affirmation or promise.

(b) Any description of the goods which is made a part of the basis of the bargain creates an express warranty that the goods shall conform to the description.

(c) Any sample or model which is made part of the basis of the bargain creates an express warranty that the whole of the goods shall conform to the sample or model.

In the sale of any product, the manufacturer or seller will describe the product in the most favorable terms frequently referred to as mere "puffing." Admittedly, the line between acceptable "puffing" and "oversell" amounting to misrepresentation is a gray area. If statements are made that exceed the performance capability of the product, liability may result.

Section 2-315 of the UCC, supporting a representation theory under an implied warranty, requires

Where the seller at the time of contracting has reason to know any particular purpose for which the goods are required and that the buyer is relying on the seller's skill or judgment to select or furnish suitable goods, there is [unless excluded or modified under the next section] an implied warranty that the goods shall be fit for such purpose.

The thrust of this language is to place a responsibility on the manufacturer or seller to supply a product meeting the purposes of the user to the extent the manufacturer or seller "has reason to know those purposes."

8.3 FACTORS RELEVANT TO DEFENDANT'S LIABILITY

As in the normal tort action for negligence, the plaintiff generally is required to establish that (1) the product was defective, (2) the defect existed at the time it left the defendant's possession, and (3) the defect was the proximate cause of the injury.

Injured Party's Conduct

Contributory negligence. In some states, when the plaintiff's product liability case is cast in terms of negligence, the defendant may successfully defeat plaintiff's recovery by proof of plaintiff's contributory negligence. Contributory negligence (discussed in detail in chapter 7) means that the plaintiff failed to act as a reasonable person and that plaintiff's conduct contributed significantly as a cause to the injury. Just as defendant's conduct is measured by the criteria of a reasonable person so is the plaintiff's conduct held to a reasonable person standard. In cases where the plaintiff fails to discover obvious defects in the product or misuses the product, the plaintiff may be found to have contributed to the injury by not acting in a reasonable manner. Of course, where the plaintiff's theory of recovery is based on the doctrine of strict liability, contributory negligence generally will not bar the plaintiff's recovery.

Assumption of risk. Plaintiff's recovery will be barred where the plaintiff's conduct amounts to an assumption of the risk. Under this defense, the defendant has the burden of proof that the plaintiff voluntarily and unreasonably accepted a known risk. On the other hand, in *Dorsey v. Yoder Company*, 331 F. Supp. 753 (E.D. Pa. 1971), the defendant failed to prove plaintiff's assumption of risk although plaintiff, a steelworker with 15 years experience, was injured when using an ungloved hand to feed sheets of steel into a rapidly moving slicing machine and a burr on one sheet caught and pulled plaintiff's hand into the machine.

Product alteration and abuse. While the manufacturer should be responsible for the products produced, if new or different components are added to the original product the issue becomes murky as to who produced the product that failed. A somewhat similar issue is raised by product abuse or misuse. While the use of a hammer cannot reasonably be expected to be limited to driving nails or pounding operations, the use of the hammer as a prying device in certain circumstances may expose the user to risks not anticipated by the manufacturer. Proof of misuse permits defendant to invoke defenses of contributory negligence or assumption of risk.

Labeling of Products

Disclaimers. Another defense available to the manufacturer is the frequent attempt to limit liability by disclaimer or by use of an express warranty of nonliability set forth in the advertisement or on the product label. One reason for legislative and judicial shifting from the implied warranty to strict liability was a desire to avoid disclaimers frequently urged by the manufacturer as part of the commercial transaction where the party injured was not a real participant in the transaction. For example, greater effect may be given to disclaimers where both parties are business entities of relatively equal bargaining strength. However, disclaimers are not likely to be given effect in consumer purchases. Section 2- 317 of the UCC permits enforcement by limiting liability through an "express" disclaimer even though inconsistent with a warranty if the disclaimer is "specific and conspicuously" disclosed.

It is common for a manufacturer to seek to limit liability for damages to the replacement of parts. The courts recognize the conflicting issues presented to the manufacturer as the sales department undertakes to convince the purchaser that the product is good ("foolproof," "unbreakable", "shockproof," etc.) and the legal department is trying to use a disclaimer to limit liability. At times, the courts find such disclaimers to be against public policy where it is clear that the buyer had no real opportunity to negotiate such terms in an ordinary retail sale.

claimers to be against public policy where it is clear that the buyer had no real opportunity to negotiate such terms in an ordinary retail sale.

Warnings. Design or manufacturing responsibility does not end with factors of functional utility. Safety in use of the product requires a total concept including the marketing elements of advertising, labels, instructions, and warnings. The manufacturer or seller assumes a duty to warn of any dangerous propensities of the product that are known or may be reasonably foreseen. The issue of whether a particular warning should be furnished is a question of what is reasonably foreseeable, and the failure to warn may be evidence of negligence if the injury is foreseeable.

An extreme illustration of a court finding that a warning label was required is found in *Moran v. Faberge*, 332 A.2d 11 (Md. 1975). A teenaged girl was injured after a friend poured the defendant manufacturer's cologne on a Christmas candle in order to "make it scented." Although no similar accident was reported in the 28 years of manufacture, the court reasoned that it was possible that some woman at her vanity table might heed a warning about inflammability if tempted to flick a cigarette ash near the bottle where the ash might squeeze through the bottle's tiny opening. This amazing holding based on possibilities as opposed to foreseeability, was rejected in *Thibault v. Sears, Roebuck & Co.*, 395 A. 2d 843 (N.H. 1978), although such warning as well as warnings that colognes and perfumes should not be stored in hot places or rubbed in the eyes now routinely appear on toiletries.

Similarly, in *Borel v. Fiberboard Paper Products Corp.*, 493 F.2d 1076 (5th Cir. 1973), plaintiff was allowed to recover for disability from asbestosis where plaintiff, working as an industrial insulator, was exposed between 1936 and 1969 to concentrations of asbestos dust and received no warning. The court's opinion noted that the plaintiff was given no option as to whether or not to risk exposure to danger since there was no warning by the defendant.

Breach of Instructions

It is customary for the manufacturer to include instructions with products describing both the proper manner of installation and maintenance as well as procedures for use and repair. Instructions tell the consumer how to use the product effectively. Warnings inform the consumer of the dangers of improper use and tell how to guard against dangers, if possible. The standard for adequate warning is indicated in *Muncy v. Magnolia Chemical Company*, 437 S.W.2d 15, (Tex. Ct. App. 1968):

of its use; the content of the warning must be of such a nature as to be
comprehensible to the average user and to convey a fair indication of the
nature and extent of the danger to the mind of a reasonably prudent person . . . the question of whether or not a given warning is legally sufficient depends upon the language used and the impression that such language is calculated to make upon the mind of the average user of the
product.

Implicit in the duty to warn is the duty to warn with a degree of intensity
that would cause a reasonable man to exercise . . . the caution commen-
surate with the potential danger . . . . A clear cautionary statement setting forth the exact nature of dangers involved would be necessary to
fully protect the seller . . . .

Since warnings are relatively inexpensive in terms of the whole
product and require no major redesign of the product, the tendency of
the manufacturers is to warn against dangers rather than redesign
against certain foreseeable dangers. The warning message should be
given to all who may have access to the product and be injured by its
use. However, practical implementation is a difficult task since injury
can occur to any persons including infants, innocent bystanders, and to
the property of absent owners.

State-of-Art Defense

In 1976, the National Machine Tool Builders Association reported
that one out of every seven lawsuits against their members involved a
machine 40 years old, one out of four involved a machine that was 30
years old, and almost all the product liability claims involved machines
more than 20 years old. Many of these machines had been sold, resold,
junked, and resurrected from the junkyard. The Illinois court in *Day v.
Barber-Coleman Co.*, 10 Ill. App. 2d 494, (Ill. 1956), dismissed a suit by
the plaintiff against the manufacturer for injury caused when an overhead door fell on him. The court reasoned,

The design having evidently been found safe in industry by experience
and having been many times used safely by the installers, the *state of the
art* at the time and prior history of the use of the product would not have
indicated or required any material change in the design or manufacture.

This defense was clarified to mean "the knowledge available both
at the time of design and manufacture." *Dean v. General Motors Corp.*,
301 F. Supp. 187, (D.C. La. 1969). In determining the technology representing the "state of the art," the criteria is not limited by what is common practice in the industry but rather by a technology forcing criteria

(i.e., "what could be done") similar to tests used under the Clean Water Act known as "best available technology" (BAT).

8.4 REMEDIES AND DAMAGES

As in ordinary tort cases, the major purpose of damages in product liability cases is to afford full compensation to the injured plaintiff. To the extent that the defendant has wrongfully taken the life, limb, or property of the plaintiff, money damages to the extent possible should undertake to restore that which was taken. Lost earnings and medical expenses may be calculated with some degree of accuracy, but, while there may be compensation for pain and suffering, establishing the appropriate value is more difficult.

Punitive damages, awarded to punish and deter a wayward defendant, are deemed an extreme remedy and normally are exacted only for flagrant, reckless violations of the victim's rights. Generally, punitive damages have been levied against perpetrators of intentional torts reflecting recklessness or gross negligence. Thus, the availability of punitive damages in product liability cases generates disagreement among commentators as the product liability theory continually ebbs toward strict liability. However, a few courts have awarded punitive damages in claims against manufacturers of defective products where the manufacturer seemed unwilling to place a high value on human life.

Because of the burgeoning cost of liability insurance and at times its complete unavailability, a number of legislative reforms of product liability law have been proposed. The proposed Uniform Products Liability Act (UPLA) adopts a strict liability standard for product defects and a negligence standard for defective design and inadequate warnings. The UPLA recognizes contributory negligence, assumption of risk, and misuse and alteration as defenses.

BEXIGA v. HAVIR MANUFACTURING CORP.
290 A.2d 281 (N.J. 1972)

Proctor, J. This is a products liability case. Plaintiff, John Bexiga, Jr., a minor, was operating a power punch press for his employer, Regina Corporation (Regina), when his right hand was crushed by the ram of the machine, resulting in the loss of fingers and deformity of his hand. His father, John Bexiga, Sr., brought this suit against Havir Manufacturing Corporation (Havir), the manufacturer of the machine, for damages in behalf of his son . . . . The action was grounded in negligence, strict liability in tort and breach of warranty of fitness of purpose. The trial court dismissed the action at the close of the plaintiffs' case. The Appellate Division affirmed and this court granted plaintiffs' petition for certification.

The machine which caused the injuries was a 10-ton punch press manufactured by Havir in 1961 and sold that same year to J.L. Lucas & Sons. Inc., a dealer, and, at its direction, shipped to Regina. With the exception of a guard over the flywheel there were no safety devices of any kind on the machine when it was shipped. Plaintiffs do not contend that the accident resulted from defective materials, workmanship or inspection. Rather, their theory is that the punch press was so dangerous in design that the manufacturer was under a duty to equip it with some form of safety device to protect the user while the machine was being operated.

[P]laintiff John Bexiga, Jr., 18 years of age and a junior in high school, had been employed by Regina at its Rahway plant for about two months, working nights after school. During his employment he operated punch presses and drilling machines for 40 hours per week.

[On the day of the accident,] June 11, 1966, John, Jr., reported for work at 5:00 P.M. and was assigned to operate a drilling machine. He worked on this machine until the 9:30 break, after which the foreman directed him to work on the Havir punch press (which he had never before operated) and instructed him in its use. Thereafter he operated the machine unattended. He testified that the punch press was approximately six or seven feet high with a ram, die and foot pedal.

The particular operation John, Jr., was directed to do required him to place round metal discs, about three inches in diameter, one at a time by hand on top of the die. Once the disk was placed on the die it was held there by the machine itself. He would then depress the foot pedal activating the machine and causing the ram to descend about five inches and punch two holes in the disc. After this operation the ram would ascend and the equipment on the press would remove the metal disc and blow the trimmings away so that the die would be clear for the next cycle. It was estimated by John, Jr., that one cycle as described above would take approximately 10 seconds and that he had completed about 270 cycles during the 40 minutes he operated the machine. He described the accident as follows:

> Well, I put the round piece of metal on the die and the metal didn't go right to the place. I was taking my hand off the machine and I noticed

that a piece of metal wasn't in place so I went right back to correct it, but at the same time, my foot had gone to the pedal, so I tried to take my hand off and jerk my foot off too and it was too late.

Plaintiff's expert, Andrew Gass, a mechanical engineer, testified that the punch press amounted to a "booby trap" because there were no safety devices in its basic design and none [was] installed prior to the accident. He added that the accident would probably never have occurred had the machine been properly designed for safety. The only literature accompanying the sale of the machine was a service manual which made no mention of safety devices in the operation of the machinery with the exception of a reference to the guard on the flywheel which was unrelated to the accident.

Gass described two "basic types" of protective safety devices both of which were known in the industry at the time of the manufacture and sale. One was a push-button device with the buttons so spaced as to require the operator to place both hands on them away from the die area to set the machine in motion. The other device was a guard rail or gate to prevent the operator's hands from entering the area between the ram and die when the machine was activated. These and other safety devices were available from companies specializing in safety equipment.

On cross-examination Gass conceded that, in accordance with the custom of the trade, presses like the one in question were not equipped with safety devices by the manufacturer. Rather, he said safety devices were to be installed by the ultimate purchaser. However, in his opinion the custom of the trade was improper in that the machine was defectively designed for safety and that purchasers "almost never" provided safety devices. Further, he said the large presses were generally equipped by the manufacturer with the push-button device. He said that smaller presses like the one in question were as dangerous to the user as the larger ones. He concluded that the press here involved should have been equipped with a two-hand push-button device as are the larger presses. On cross-examination he was not asked to explain why push-button devices were installed by the manufacturer on the large presses but not on the small ones.

While pointing out that guard rails or gates might have to be "modified" to suit the particular die or part used with the press, Gass stated that the push-button device would not have to be "modified" no matter what die was used. In other words, the push-button device would be appropriate for any of the machine's normal uses. On cross-examination he admitted, if the press was employed to punch holes in a four-foot pipe, a guard rail or gate would impede entry of the pipe into the die area and would have to be removed. He said that in such a case the guard rail or gate would not be needed because in holding the pipe the operator would be standing away from the machine.

The Appellate Division in affirming the trial court's dismissal held that plaintiffs failed to make out a prima facie case under strict liability, breach of warranty or negligence principles . . . . The court reasoned that since it was the custom of the trade that purchasers, rather than manufacturers, provide safety devices on punch presses like the one in question, Havir "had no reason to believe that the press would be put to use without some additions, i.e., the installation by Regina of protective devices suitable to whatever manufacturing process the press was to be devoted." It also stated that N.J.S.A. 34:6-62, in effect at the time of the sale, required the factory owner to equip its power presses with proper guards. It held liability could not be imposed under the RESTATEMENT rule because the manufacturer did not expect the product to reach the user without substantial change.

Taking as true all evidence supporting plaintiffs' position and according them the benefit of all inferences which can reasonably be drawn therefrom, . . . [w]e cannot agree with the Appellate Division's application of the law on either negligence or strict liability on the evidence presented. We have concluded that on either theory the proofs were sufficient to withstand a motion for dismissal.

There is no question but that the punch press here without any safety devices was dangerous to the user. From the evidence as to the guard rails or gates mentioned above we agree with the Appellate Division that it would be impracticable for the manufacturer to equip his presses with all of these protective devices, and therefore improper to place the responsibility for their installation upon the defendant. However, the expert testified that the alternative basic safety device, the push- button guard, would not have to be "modified" to suit the die press and there was no evidence that that device would have to be changed for any of the varied uses of the machine.

On the basis of Gass' testimony the jury could infer that the two-hand device was appropriate for every normal operation of the machine and, thus, that it was not impracticable for Havir to equip its machine with such a device. Moreover, as noted above, the expert pointed out that larger punch presses are equipped by the manufacturer with push-button devices; that smaller presses, such as the one in question, were just as dangerous as the larger ones; and that they should be prepared for safety in the same manner. The jury could infer that the two-hand device was appropriate for all of the operations performed on the larger presses and, since the expert was not cross-examined as to why there should be a difference as to the safety devices provided on the two types of presses, it could also infer that the push-button device was equally appropriate for the normal uses of the smaller presses.

As we previously said on the issue of strict liability, the Appellate Division applied the rule set forth in the RESTATEMENT. To the extent that the rule absolves the manufacturer of liability where he may expect the purchaser to provide safety devices, it should not be applied. Where a manufacturer places into the channels of trade a finished product which can be put to use and which should be provided with safety devices because without such it creates an unreasonable risk of harm, and where such safety devices can feasibly be installed by the manufacturer, the fact that he expects that someone else will install such devices should not immunize him. The public interest in assuring that safety devices are installed demands more from the manufacturer than to permit him to leave such a critical phase of his manufacturing process to the haphazard conduct of the ultimate purchaser. The only way to be certain that such devices will be installed on all machines—which clearly the public interest requires—is to place the duty on the manufacturer where it is feasible for him to do so.

We hold that where there is an unreasonable risk of harm to the user of a machine which has no protective safety device, as here, the jury may infer that the machine was defective in design unless it finds that the incorporation by the manufacturer of a safety device would render the machine unusable for its intended purposes. As we have said, the jury could infer from plaintiffs' evidence that it was feasible for Havir to install the push-button device. Therefore, it was error for the trial court to dismiss the strict liability claim at the close of plaintiffs' case.

Of course the question of whether one is negligent depends on whether he acted reasonably under the circumstances of a particular case. Thus, aside from the question of the practicability of Havir's installation of a safety device on the press here involved as to negligence, we must consider whether Havir could reasonably foresee that Regina would not install a safety device. On this issue the custom of the trade—that the manufacturer did not install safety devices but relied on purchaser to do so—while ordinarily evidential, is not conclusive. Nor would it be conclusive that [the state law] imposed the duty on the purchaser. While Havir may have thought that Regina would have taken adequate precautions to protect its employees or that it would be required to do so by statute, we do not think in view of the circumstances here that Havir had a right as a matter of law to assume such devices would be provided. As to negligence, we hold that a jury question was presented, and that it was error for the trial court to dismiss the action at the close of plaintiffs' case.

If the jury should find for the defendant on the issue of the defective or negligent design of the machine, the next question for it should be whether the defendant was negligent in failing to attach to the machine a suitable warning to the operator of the danger of using it without a protective device.

Because of our disposition of the case it is necessary to consider the defendant's contention that John, Jr., was contributorily negligent as a matter of law. Neither court below decided this issue. Contributory negligence may be a defense to a strict liability action as well as to a negligence action. However, in negligence cases the defense has been held to be unavailable where considerations of policy and justice dictate. This court said that undoubtedly the defense will be unavailable in special situations within the strict liability field. We think this case presents a situation where the interests of justice dictate that contributory negligence be unavailable as a defense to either negligence or strict liability claims.

The asserted negligence of plaintiff—placing his hand under the ram while at the same time depressing the foot pedal—was the very eventuality the safety devices were designed to guard against. It would be anomalous to hold that defendant has a duty to install safety devices but a breach of that duty results in no liability for the very injury the duty was meant to protect against. We hold that under the facts presented to us in this case the defense of contributory negligence is unavailable.

The judgment of the Appellate Division is reversed and the cause is remanded for a new trial.

SKYHOOK CORPORATION v. JASPER
560 P.2d 934, (N.M. 1977)

Oman, Chief Justice. This is an action for claimed wrongful death brought by plaintiff (Jasper), as administrator of the estate and personal representative of Melvin Mack Brown, deceased. Decedent was employed by Electrical Products Signs, Inc. (Signs, Inc.) as an apprentice sign installer. On January 11, 1973, he was assisting a journeyman installer of signs (Pulis), also employed by Signs, Inc., to install a Phillips 66 sign at a service station near Springer, New Mexico.

A hole had been dug in the ground in which to place the heavy signpost, a metal pipe, in an upright position. Pulis and decedent were using a 100 foot telescoping crane rig to lift and place the signpost in the hole. This crane was manufactured by Skyhook and sold by it to Signs, Inc., in January 1968. A clearly visible written warning appeared on the boom. In this warning it was stated: "All equipment shall be so positioned, equipped or protected so no part shall be capable of coming within ten feet of high voltage lines."

Pulis was aware of and had read the warning, and the evidence is to the effect that decedent also had seen and was aware of the warning, since it was clearly visible and decedent had previously worked on and had operated the rig. Both Pulis and decedent knew of the presence of overhead high voltage lines, since they had been warned of the presence of these lines by the operator of the Phillips 66 station at which the sign was being installed. The station operator had warned them that they should not operate the equipment ten feet from these high voltage lines.

Pulis and decedent positioned the crane so that, in the judgment of Pulis, the crane was ten or twelve feet from the power lines. However, no measurements were made to assure that the positioned distance of the crane from the power lines was sufficient to prevent any portion of the equipment from coming within ten feet of these lines, even though a tape measure was kept in the cab of the rig for the purpose of making these measurements. Pulis then hoisted the signpost with the crane and began swinging it toward the hole in which it was to be positioned. As he was swinging the signpost toward the hole, he heard decedent scream. Decedent, who was guiding the signpost by hand toward the hole, was electrocuted when the lift cable came in contact with the overhead power line. A "tag line" or "guide rope," which was not an effective conductor of electricity and which decedent could have used to guide the signpost to the hole, was available, but was not ordinarily used by the helper in setting a post. There were also other measures commonly known, and known at least to Pulis, which could have been taken to avert the electrocution of decedent.

Decedent had been warned by his father of the dangers in operating a crane too near high voltage lines. The rig had been used by Signs, Inc., for the purpose of erecting signs for a period of five years, and no such accident or incident had ever previously occurred . . . .

Plaintiff sought recovery from Skyhook on the theory of strict tort liability for failing to equip its crane, at the time of its sale to Signs, Inc., in January 1968, with either an "insulated link" or a "proximity warning device." An insulated link is a device installed on a crane to isolate the lifting hook from the lifting line or cable, so that there is no electrical continuity between the crane boom or lifting cable and the load being lifted. In January 1968, no crane manufacturer installed insulated links as standard equipment, but they were available to a purchaser of a crane at an additional cost of $300 to $400, depending on the size of the link.

A proximity warning device is an alarm warning system activated by the electrostatic field of overhead power lines. The use of this device requires that the crane be positioned at the minimum distance desired from the power line and the device then set for operation. If properly set, it will warn the operator by sound and lights when the equipment encroaches on the minimum pre-set distance from the power line. At the time of the sale of the crane to Signs, Inc., no crane manufacturer offered this device as either standard or optional equipment, but it could be purchased for approximately $700 . . . .

[T]he question to be resolved is whether the evidence created an issue of fact as to liability of Skyhook under Section 402A [of UCC], which should have been submitted to the jury. . . .

There is no question about the sale of the rig by Skyhook to Signs, Inc.; no question that Skyhook was engaged in the business of selling these rigs; no question that decedent was using the crane rig at the time of his death; and no question about any substantial change having been effected in the rig from the day of its sale to Signs, Inc., in January 1968 to January 11, 1973, the date of the unfortunate accident.

Therefore, the only issue under Section 402A which must be determined [is] whether the crane was in a defective condition which made it unreasonably dangerous to the user.

First, we must decide whether the failure of a seller to include an optional safety device as a part of the product may be considered as a sale of the product in a "defective condition." It would serve no useful purpose to try to reconcile the authorities on this point. However, we are of the opinion that a failure to incorporate into a product a safety feature or device may constitute a defective condition of the product. Obviously, the test of whether or not such a failure constitutes a defect is whether the product, absent such feature or device, is unreasonably dangerous to the user or consumer or his property.

The crane rig had been used by Signs, Inc., for five years, had performed well, and no injury had resulted. Obviously, it was not unreasonably dangerous within the contemplation of the ordinary consumer or user of such a rig when used in the ordinary ways and for the ordinary purposes for which such a rig is used. Furthermore, even though Skyhook had knowledge that the rig might be used in areas where overhead high voltage lines were present, it placed on the boom a clearly visible written warning that "all equipment shall be so positioned, equipped or protected so that no part shall be capable of coming within ten feet of high voltage lines." There is no contention that this warning was inadequate, had it been heeded. Skyhook, as the seller, could reasonably assume that the warning would be read and heeded.

And had it been heeded, the crane rig was not in a defective condition nor unreasonably dangerous.

The above reasons are sufficient in themselves to dispose of this case, but we have more here. Both Pulis and decedent had the presence of the high voltage lines called to their attention, both knew the dangers of high voltage electricity—as does every ordinary adult in this present day society in New Mexico in which electricity is used so commonly for so many purposes—and together they positioned the crane rig away—but not far enough away—from these high voltage lines. There is no duty to warn of dangers actually known to the user of a product, regardless of whether the duty rests in negligence or on strict liability under Section 402A.

Since there was no defect in the crane rig unreasonably dangerous to the decedent within the contemplation of the strict liability concept enunciated in Section 402A, there was no culpable conduct on the part of Skyhook which could have proximately caused the accident and the resulting death.

The decision of the Court of Appeals is reversed and this cause is remanded to that court with directions to affirm the judgment of the district court. [For defendant]

DISCUSSION QUESTIONS

1. If a manufacturer's product meets the "state of the art" at the time of manufacture, what is the subsequent legal responsibility of the manufacturer to retrofit earlier versions of the product when there is a drastic safety improvement in the state of the art? Could a safety-conscious manufacturer ethically justify not making such retrofits?

2. Discuss the responsibility of a manufacturer in connection with the unintended use of its product (e.g., a stepladder used as bridge rather than a plank). If misuse of the product is foreseeable, does this raise a duty to issue a warning against such misuses (e.g., the manufacturer of stepladders warn against standing on the top step or redesigning the top step so it can not be used as a step)?

3. In a case involving an allergic reaction to a product, the court suggests a two part test for imposing a manufacturer's duty to warn (1) that the injured person be a member of a "substantial number" of persons injured by the product and (2) that the manufacturer knew or should have known of this class of persons. What is the meaning of a "substantial number," and why does the court set down such a requirement?

4. Discuss the social and economic implications of a trend toward the doctrine of "strict liability" rather than the traditional doctrine of "negligence" in product liability cases?

5. Illustrate the meaning of a product which is "unreasonably dangerous" under Section 402A of the RESTATEMENT. Does this affect the burden of proof placed on the injured plaintiff?

6. If the court uses the ability of the manufacturer to obtain insurance as a basis of liability, what are the economic and safety policy implications?

7. How can the *Bexiga case* and the *Skyhook case* be reconciled?

9

AGENCY AND BUSINESS ENTERPRISES

9.1 AGENCY: CREATION AND AUTHORITY

Nature of Agency

In the modern commercial enterprise it is essential that operations be conducted through structured organizations of various persons since face-to-face dealing of the principals either from the beginning to end of negotiations or on all aspects of a project is inefficient if not impossible. To fill this gap, an agency arrangement is frequently used, and that relationship, a consensual affiliation between two persons or entities, involves a situation under which the first person, the principal, undertakes to act through a second person, the agent, with respect to a transaction with a third person. In a sense, the arrangement may be described as "let Jack do it," where the agent agrees to act for the benefit of and under the control of the principal. The agent is not a mere volunteer but operates on behalf of and under the authority granted by the principal. The most common questions that arise are (1) whether the principal is contractually bound by the acts of the agent and (2) whether the principal is liable for the wrongful (tortious and criminal) acts of the agent.

Agents are subclassified as (1) general agent or (2) special agent, with the distinction being that the general agent normally conducts a series of transactions on behalf of the principal over a period of time while the special agent may be authorized to conduct more specific acts or a single transaction for a limited time. The former is illustrated by an insurance agent and the latter by a real estate broker handling a single property sale.

In the familiar employer/employee relationship, the employer exercises considerable control over the physical conduct of the employee during so-called working hours. Thus, building craftspeople or factory workers perform under the instructions and direction of a supervisor. In the principal/agent relationship, the principal has control over the agent, but the control generally relates to specific business activities of the agent such as the functions of a real estate broker who normally does not have a fixed working schedule.

The independent contractor relationship on the surface closely resembles the principal/agent and the employer/employee relationship, but must be differentiated from the other arrangements. While the "independent contractor" has many of the indicia of the other two situations, the legal status of the independent contractor is different and is covered by a separate and distinct body of law. Absent special circumstances (e.g., handling of extra hazardous or dangerous materials), the person using the independent contractor relationship is not contractually bound by the independent contractor's acts or responsible for the independent contractor's wrongful tortious or criminal conduct. In *Karnowski v. Skelly Oil Co.*, 174 F.2d 770 (10th Cir. 1949), plaintiff was injured by a roofing stone thrown off the roof by the repairman who had been engaged to make roof repairs by defendant building owner, and was denied recovery since the repairman was held to be an independent contractor.

Whether an individual or entity falls within the status of an independent contractor is determined by a number of factors that indicate the degree of control exercised over the operations. A high degree of control and direction points toward an employer/employee or principal/agent relationship. On the other hand, an independent contractor agrees to do a project according to the contractor's own methods and is obligated only for the final work product while retaining the right to exercise independent judgment concerning the manner in which to proceed toward that objective. The ultimate determination of the relationship is critical in holding or dismissing the alleged employer/principal from liability for the conduct of the independent contractor.

In *Massey v. Tube Art Display, Inc.*, 551 P.2d 1387 (Ct. App. Wash. 1976), the court found ample evidence of control to reject the defendant's contention that Redford, the owner and operator of a backhoe, was an independent contractor and found that an employer/employee status existed. There an individual, Redford, spent approximately 90% of his working time digging holes for the defendant sign company. Although Redford had no employees and paid his own business taxes, Redford did not obtain the permits for the jobs and simply dug holes to a designated depth at locations marked by yellow paint. One evening, while digging a hole, Redford hit a small natural gas pipeline, but, after examining it, concluded that it was not leaking and left it for the night. Later that night there was an explosion and fire, which damaged plaintiff's adjoining building. The court permitted plaintiff to recover from defendant, finding Redford an employee based on the extent of defendant's control over Redford's work.

Creation of Agency Relationship: Capacity

An agency arrangement is created by act of the parties or by operation of law. Generally, no formality is required for the creation of the agency relationship, which may be created by either express consent or by implied conduct. Under certain statutes such as the Statute of Frauds, written evidence of the relationship may be required if it is to be effective for certain transactions such as those involving real property. Since a requisite to the creation of an agency relationship is the appropriate capacity of the parties, the principal must have contractual capacity and the right to enter into a contract. It follows that a minor who does not have legal capacity to enter into a contract cannot be a principal in a binding contract by means of an agency relationship. On the other hand, a minor, subject to having a minimum intellectual capacity (i.e., ability to recognize fiduciary duties), may serve validly as an agent on the grounds that it is the principal who is bound by the contract and not the agent. In certain circumstances, a person may be disqualified from being an agent because of a conflict of interest by representing both parties, dealing secretly for the agent's own account, or acting without an appropriate license.

Agent's Authority to Bind Principal

The authority of the agent is a key issue with respect to whether the principal is contractually bound by the agent's acts. The authority of an agent may be either actual or apparent, and that authority is conferred through some writing, words, or conduct of the principal making it clear that the principal intends the agent to so act.

Actual authority. Actual authority is the authority that the principal intentionally delegates to the agent. Actual authority may flow from express statements or may be implied as incidentally flowing from the express authority, from either custom or usage, by acquiescence of the principal or by emergency or necessity. A unilateral act of the principal is sufficient to create express authority if it reasonably leads the agent to believe that the principal wants the agent to act on the principal's behalf. The conclusive nature of express grants of authority is demonstrated by *Allen A. Funt Productions v. Chemical Bank*, 405 N.Y.S. 2d 94 (1978), where the court held the defendant bank was entitled to rely on the express authority for checks written and signed by plaintiff Funt's accountant even though such checks were a part of the accountant's embezzlement scheme. The court found that the bank was not negligent since plaintiff Funt had expressly authorized the accountant to sign checks and the amount of the checks were within the authority limits.

Scandinavian law, which has been described as a model for a uniform international law with respect to apparent authority of the agent and the proposed draft of the Rome Institute for Unification of Private Law, although cast in terms of implied authority provided,

> Article 4—*Implied Authorization.* The authority of a person to act in the name of another may arise from some position which that person occupies with the consent of the other, and from which the power to act in the name of the other arises according to the law and usages applicable.

> Article 8—*Scope of Authorization Implied from a Position.* In the case of authorization implied from a position, the agent shall be authorized to perform in the name of his principal all those acts normally implied by his position.

Implied authority can be construed as actual authority derived from the principal's explicit words or conduct and which give the agent power to do those things that are usual and incidental to the express authority, reasonably necessary to achieve the goals assigned by the principal.

The German law presents an interesting comparison between different legal systems in which agency is referred to as "procura." The practice is for a commercial employer to give a document (called the procura) to an individual, but it must contain the word "Prokura" and be registered in the Commercial Register. The distinctive feature of the procura is that it gives to the procurist (agent) power to bind the principal in all transactions except real estate. While not restricted to the usual scope of business, the power has only two limitations. First, it may be exercised only for a particular branch of business, located in a separate city, and, second, that the third party cannot hold the principal if the third party dealt in bad faith with the procurist.

Apparent authority. Apparent authority or ostensible authority, as it is sometimes called, exists where the principal "holds out" another, the agent, as possessing certain authority, inducing others reasonably to believe that authority exists. It is that semblance of authority that the principal's acts or inadvertences cause or allow third persons to believe the agent possesses. The agent has apparent authority to act even though, as between the principal and the agent, such authority has not been expressly granted. In the case of apparent authority, the authority is created by the acts or conduct of the principal "communicated to the third party" as opposed to actual authority, which is "communicated by the principal to the agent." The "holding out" by the principal may be conveyed either by affirmative acts or by inaction.

In *Clark Advertising Agency v. Tice*, 490 F.2d 834 (5th Cir. 1974), where the president of the National Hot Rod Association (NHRD) left the negotiations of a contract to the NHRD's vice president and controller, the court held that this officer had apparent authority to bind NHRA.

Ratification. Ratification is the subsequent acceptance or approval by one person, the principal, of an unauthorized act of another person, the agent, and refers to the time of the original action by the agent. In such situations, although the agent's action was beyond the scope of the agent's actual, implied, or apparent authority, the principal nevertheless may become bound as if the agent's action was originally authorized. Generally, if the agent's act could have been legally performed or authorized by the principal, the principal may ratify the agent's earlier unauthorized action, and new consent by the third party is not required. Ratification may be express or implied. Express ratification is indicated by affirmative approval of the agent's earlier unauthorized action, while implied ratification results from accepting the benefits of the transaction or may be ratified by silence where the principal has a duty to disaffirm the transaction. For example, if the principal brings a lawsuit involving the transaction, that action will be construed as a ratification. Generally, ratification will not be implied unless the principal has or reasonably should have knowledge of the material facts concerning the transaction at the time of the principal's affirmation of the agent's acts.

Four elements required for valid ratification follow: (1) The agent's action must have been a valid act at the time it was performed, (2) the purported principal/agent relationship must have existed and the agent must have acted on behalf of the principal, (3) the ratification must have been accomplished with any required formality, and (4) the principal must know the material facts at the time of ratification.

In ratifying the transaction, the principal must accept the entire transaction and may not ratify merely a portion of the transaction except for acts associated but clearly a separate and severable transaction. There is no essential difference between the capacity to appoint an agent and the capacity to ratify unauthorized action by an agent. In *Linn v. Kendall*, 238 N.W. 547 (Iowa 1931), an owner was held liable to the architect for drawing plans where the contractor requested the architect to draw plans that the owner later approved.

9.2 CONTRACTUAL RELATIONSHIP OF PRINCIPAL AND THIRD PARTY

Disclosed Principal

A disclosed principal represents the usual agency relationship and is, as the name implies, a principal whose existence and identity are known to the third parties. Normally, the agent of a disclosed principal (when acting within the scope of the agent's authority) does not intend that the agent will become a party to the transaction. Customarily, the instruments are drawn in the name of the principal, indicating that the agent is acting as agent for the principal except in unusual situations when the parties intend that the agent also be a party to the contract.

Undisclosed Principal

The law has been prone to hold persons who exchanged promises in commercial transactions to their obligations. Unfortunately, this concept and its consequence are not always uppermost in the minds of persons who carry out negotiations to buy lumber or steel or to undertake preliminaries of construction. Consequently, the agreement may require "that certain material be on the job site by the 10th" but may not always clearly specify for whose benefit the transaction is being made. The contractor or purchasing agent may intend these details to be sorted out by the bookkeepers so that charges may be allocated to the correct job, but the purchase orders on their face may not identify the parties fully.

An undisclosed principal is one whose existence and identity are unknown to the third party. A partially disclosed principal is one whose existence but not identity is known to the third party. Both the agent and the principal are liable on a contract entered into by an authorized agent on behalf of a partially disclosed or undisclosed principal. The third party, upon learning the existence and identity of the principal, must decide whom the third party intends to hold liable. The third party may not hold both the agent and the principal liable.

While only the principal and not the agent is entitled to enforce a contract in which the principal is disclosed, both the principal and the agent may enforce the contract where the principal is partially disclosed or the principal is undisclosed. It is irrelevant that the third party did not know of the existence of a principal and only knew of the agent.

There are certain cases in which the principal may not enforce the contract against the third party. First, where the agent has fraudulently concealed the identity of the principal, the courts will not order specific performance against the third party and will permit the third party to rescind the contract. The courts reason that enforcement would not be proper where the agent's fraudulent concealment constitutes an affirmative misrepresentation. Second, where performance will impose an undue burden on the third party (such as a requirements contract or contracts involving personal trust, credit, or confidence), the contract may not be specifically enforced and the third party may rescind the contract.

9.3 CONTRACTUAL RELATIONSHIP OF AGENT AND THIRD PARTY

An agent, acting within the scope of the agent's authority on behalf of a disclosed principal, is not liable on contracts unless the agent expressly indicates an intent to be liable. On the other hand, as indicated in the previous section, the agent is responsible at the third party's election on contracts executed on behalf of a partially or undisclosed principal.

In addition, when the agent has no authority or exceeds authority, the agent is liable to the third party for breach of an implied warranty of authority. Where the authority is intentionally misrepresented, the purported agent will be liable for damages in a tort action for deceit.

9.4 CONTRACTUAL RELATIONSHIP OF PRINCIPAL TO AGENT

Duties of Agent to Principal

The agent's duties flow primarily from the contractual arrangement, if any, with the principal. In addition, the agent has three major responsibilities implied by law: loyalty, obedience, and care.

Duty of loyalty. The agent owes a fiduciary duty to the principal of undivided loyalty similar to the fiduciary duty of a trustee. An agent must act exclusively for the benefit of the principal and breaches the duty of loyalty by having interests adverse to the interests of the principal (e.g., self dealing or obtaining secret profits). An agent owes a duty to the principal not to use confidential information obtained in the course of the agency for the agent's own benefit to the detriment of the principal. While the agent cannot use specialized information for individual profit, agents who acquire generalized information during the course of employment may use that general knowledge. However, under normal circumstances, after the agency is terminated, the agent may engage in the same business unless restricted by agreement. The courts are reluctant to enforce agreements not to compete unless restricted as to a reasonable time and reasonable area, but the agent may not use or reveal confidential information acquired during the agency to the detriment of the principal. In *Hartung v. Architects Hartung/Odle/Burke Inc.*, 302 N.E.2d 240 (Ind. App. 1973), defendant architect, a former proprietor in an incorporated architectural business, was held liable for breach of fiduciary duties to the corporation and the other principals when the defendant architect resigned and took sole possession of offices and contracts of the corporation. Under the rubric of loyalty, the agent is responsible to the principal to account for any money or property that comes into the agent's possession while transacting the principal's business. Likewise, the agent is under a duty not to commingle the money or property of the principal with the agent's own property or money.

Duty of obedience. The principal has the right to instruct the agent with respect to how the agent should perform the service, and the agent is obligated to follow all "reasonable" directions of the principal. The agent has the duty of good conduct and is responsible for losses suffered by the principal resulting from the failure to follow instructions; however, the agent is not obligated to perform any act that is illegal or unethical. If an emergency occurs, the agent may disregard previous instructions where in the agent's good faith judgment to follow them would result in injury to the principal.

Duty of reasonable care. An agent owes a duty to the principal to carry out the agency duties with reasonable care considering local community standards and taking into account any special skills of the agent.

Remedies of the Principal

Where the agent's conduct injures the principal, the agent may be liable either for breach of contract or for wrongful tortious conduct. The agent's liability includes damages suffered by the principal as the result of breach of contract and for damages to the principal's property resulting from misuse of the principal's property, whether the agent's misuse was due to intentional or negligent performance or failure to perform. Where the agent breaches a fiduciary duty and obtains secret profits, the principal as in other intentional breaches may withhold the agent's compensation in addition to recovering the actual profits or property held by the agent.

Duties of Principal to Agent and Remedies of Agent

Since an agency relationship envisions the agent as an active participant in the implementation of the transaction (i.e., negotiation), the principal's instructions are skewed toward describing the agent's conduct; however, the RESTATEMENT (Agency) Section 434 mandates that the principal has the duty "to refrain from unreasonably interfering with [the agent's] work." The contract, between the principal and agent, usually specifies the compensation of the agent; but if there is no express arrangement for compensation, the principal has a duty to compensate the agent reasonably for the agent's services unless the agent is committed to act gratuitously. The scope of this payment to the agent should be sufficient to reimburse the agent for all expenses and losses reasonably incurred including legal expenses in discharging authorized duties. In some jurisdictions, the agent has a possessory lien against principal's property held by agent for money due agent.

9.5 TORT LIABILITY: RESPONDEAT SUPERIOR

While the rules for imposing tort liability on the principal for the acts of the agent are more limited in scope than those concerning contract liability, a principal may be liable to third parties for torts committed by the agent under the doctrine of *respondeat superior* ("let the master answer") for a wrongful act committed within the scope of the employment relationship. The subject of vicarious liability of the principal/employer for injuries caused by the torts of the agent/employee was discussed in more detail in chapter 7.

The liability of the principal/employer for the torts of the agent/employee is joint and several inasmuch as the injured third party may recover from either the principal/employer or the agent/employee. Since the liability of the principal/employer is derivative (i.e., the principal is answering for the wrongful conduct of the agent/employee), a release or determination of nonliability as to agent/employee generally precludes recovery from the principal/employer. In considering whether the injury was within the agent/employee's scope of employment, three tests are commonly considered: (1) whether the conduct was of the same general nature as or incidental to that for which the agent was employed to perform, (2) whether the conduct was substantially removed from the authorized time and space limits of the employment, and (3) whether the conduct was actuated at least in part by a purpose to serve the principal. Under these rules, the principal/employer is not generally liable for the intentional torts of the agents on the simple ground that an agent is not employed to commit intentional torts. However, where the intentional tort occurs as a natural consequence of carrying out the principal/employer's business, the courts have found the principal liable. Security guards, bouncers, and bill collectors illustrate areas of employment where the nature of the employment naturally engenders friction or authorizes the use of force.

9.6 PARTNERSHIP: NATURE AND FORMATION

In General

The law of partnership embraces rules that had their origin in the Civil law, the Common Law, equity, and the law merchant. There is evidence of legal regulation of partners as early as 2300 B.C. under the Code of Hammurabi. Today, the law of partnerships, based upon statutory enactment of the Uniform Partnership Act (UPA), adopted in most states, draws heavily on the law of contracts and agency. The UPA defines a partnership as "an association of two or more persons, to carry on business for profit, as co-owners." Since either an express or implied contract is essential to the formation of every partnership, individual partners must have the capacity to enter into a contract. While no particular formality is essential, partnership agreements that fall under the Statute of Frauds are required to be in writing. As with other contracts, the purpose for which a partnership is formed must be legal.

Where there is no ascertainable form of agreement, the courts generally look to the intent of the parties to determine the existence of a partnership. Under UPA, the courts evaluate the evidence (giving no single element conclusiveness) and consider certain acts of the parties to establish the nature of the arrangement relating to (1) joint or common tenancies of property, (2) designation of entity as "partnership," (3) sharing of gross returns, and (4) sharing of profits/losses.

Written Agreement

A properly drafted partnership agreement containing numerous carefully prepared provisions tailored to the particular partnership arrangement is desirable. Clauses frequently referred to as "boiler plate" (i.e., clauses that are common to most partnership agreements or those that usually appear in printed forms) will identify the partners with their addresses, the name of the partnership, and its duration as well as the nature and place of business. Specific provisions will set forth the capital or other contributions by the partners, the respective duties of the partners, a method of settling disputes, matters concerning profits, the withdrawal of funds, and the means for dissolution of the partnership.

Partnership Name

Most partnerships operate under a partnership name of one or more of the partners or some fictitious name. Many jurisdictions place some limitations on the names so that they cannot include the words "company," "incorporated" or "limited." Generally, all jurisdictions have fictitious name statutes that permit the selection and use of a name and the filing of that name with certain county or state offices. Examples are Jones and Miller, Aero Electric, and Smith d/b/a/ Enterprise Construction.

9.7   RELATIONSHIP AMONG PARTNERS

Partnership Relationship

The partnership relationship imposes fiduciary duties on all partners comparable to those obligations in the principal-agent relationship. Section 21(1) of UPA provides

> Every partner must account to the partnership for any benefit, and hold as trustee for it any profits derived by him without the consent of the other partners from any transaction connected with the formation, conduct or liquidation of the partnership or from any use by him of its property.

Section 8 of the UPA provides a sufficiently broad definition of partnership property to embrace everything that the partnership owns, including all property originally brought into the partnership stock plus other assets subsequently acquired by purchase or otherwise on account of the partnership. The intention of the partners is generally control-ling as to what is partnership property and what remains their own in-dividual property. However, unless clearly indicated to the contrary, all property acquired with partnership funds becomes partnership assets. The distinction between partnership and individual property is sig-nificant for various reasons including tax purposes and in the event of insolvency.

The partnership, as contrasted with the corporate form, is not a separate entity as a corporation and does not have a separate existence from its partners. Thus, a change in the membership of the partner-ship, either by a partner ceasing to be a member or the addition of a new partner, technically dissolves the existing partnership and requires the formation of a new partnership.

Management of Partnership Business

Unless there is an agreement to the contrary, all partners are en-titled to an equal voice in the management of the partnership business. While this doctrine of equality provides for equal participation in the management rather than being determined by the respective share (percentage of partnership ownership) of each partner in the business, it is customary in partnership agreements to modify this doctrine of equality to provide that the voice and control of the various partners will be determined by their capital contribution. Absent a stipulation in the partnership arrangement to the contrary, a majority vote of the partnership interests will control ordinary matters arising in the con-duct of the partnership business but it should be noted that Section 18(h) of UPA provides that no act that contravenes the partnership agreement may be taken without the consent of all partners.

In the partner relationship, each partner has the capacity to bind the other partners by acts and representations ostensibly within the scope of the partnership business. Section 9 (1) of the Partnership Act provides

> Every partner is an agent of the partnership for the purpose of its busi-ness, and the act of every partner, including the execution in the partner-ship name of any instrument, for apparently carrying on in the usual way the business of the partnership of which he is a member binds the partnership, unless the partner so acting has in fact no authority to act for the partnership in the particular matter, and the person with whom he is dealing has knowledge of the fact that he has no such authority.

While the articles of partnership may place explicit limitation on the powers of certain members of the partnership, such restrictions are ineffective with respect to third parties who deal with a partner without knowledge of such limitation.

Profits and Losses: Liability of Partners

An essential feature of a partnership is the right to share in the profits and the correlative obligation to share in the losses. Profits and losses may be shared pursuant to any formula upon which the partners agree.

Tort liability. Responsibility for a partnership tort is joint and several with the result that every member of the partnership is individually liable as well as all of the partners jointly. Recovery against an individual partner is not dependent on the personal wrong of that particular partner. Liability flows from any tort committed by any partner or employee of the partnership in the course of partnership business. This requires a determination similar to the determination of "scope of employment" and depends on whether the tort was committed within the ordinary course of the partnership business as distinguished from the partner's individual or personal purposes.

Contract liability. Since all partners are liable for the debts and obligations of the partnership, individual partners as well as the partnership firm are liable for the debts and obligations of the partnership. While the partnership is not a legal entity separate from its partners, statutes in most jurisdictions now provide that persons carrying on business as a partnership may sue or be sued in the partnership name.

9.8 CORPORATION: DEFINITION AND FORMATION

Corporation: Nature of Entity and the Pros and Cons

A corporation is a separate legal entity having an existence distinct from that of its owners (referred to as shareholders or stockholders) or those who manage its affairs (officers and directors). The corporate entity comes into existence by charter from the state. What follows are advantages of the corporate form of doing business: (1) shareholders have limited liability for corporate debt and obligations, (2) shareholders ordinarily may freely transfer their ownership of shares in the corporation to others, (3) the corporation has a continuity of existence that may be perpetual and its existence is not affected by a change in ownership, (4) the form of management decision making is regularized by statute and the by-laws, and (5) access to capital formation because of the comparative ease with which shares in the corporation may be bought and sold. There are also disadvantages: (1) formalities of corporate organization are more elaborate and expensive, (2) the corporation is a separate tax paying entity for income tax purposes and taxes may be incurred both when the corporation realizes income and again when dividends are distributed to the shareholders, and (3) formalities of management may be more restrictive.

Purposes and Powers of the Corporation

The purposes for which the corporation are formed are generally set out in the articles of incorporation. Although under earlier corporate law, the purposes of the corporation were set forth with considerable specificity (such as manufacturing steel, engaging in road building, or exploring and producing petroleum), the modern trend is to describe the scope of the corporation as "for any lawful purpose."

The corporation's powers are the means used to accomplish its purposes. In addition to the express powers of the corporation to perform any act specifically authorized by statute or its articles of incorporation, the corporation has implied powers to perform all other acts "reasonably necessary" to accomplish its purposes and not otherwise unlawful or prohibited.

Problem areas with respect to corporate powers that arose under earlier cases involved (1) guarantees, (2) participation in a partnership, and (3) donations. Under the modern statutes, a corporation may guarantee loans of others when such action is "in furtherance of the corporate business such as guaranteeing a loan to a customer to enable the customer to buy the corporation's product." With respect to the argument that a corporation should not join a partnership, the rationale of the cases was that, under established principles of partnership, the corporation would be bound by the acts of its copartners, who were not duly appointed agents or officers of the corporation, and that such unrestricted blanket responsibility could not be delegated by the board of directors. While a number of older cases precluded the donation of corporate funds, many modern statutes expressly grant power to corporations to make donations to humanitarian, educational, philanthropic, or other public activities. The underlying reasoning for the earlier cases was that the corporation was established to conduct business activity with a sole view of corporate profit and the charitable contributions did not further that objective. Later cases dropped the requirement of "direct corporate benefit" and looked more to the reasonableness of the donation on the grounds that activity that "maintains a healthy social system" necessarily serves a long run corporate purpose. As an extension of this line of thinking, it is commonplace for shareholders and special interest groups to exert pressure on corporations to achieve sociopolitical goals (e.g., to end apartheid and refrain from the manufacture of napalm).

Formation of Corporation

State statutes set forth the requirements for the valid formation of a corporation. Generally, the states specify the number of incorporators who are required to prepare and file the articles of incorporation (referred to in some states as "certificate of incorporation" or "charter"). This document can be very simple and, for example, provide for a corporate name, principal office, purpose of the corporation, description of the capital structure, and size of the board of directors. When the articles of incorporation are filed properly together with the appropriate fee, the corporation is ready to commence business. At this stage, a board of directors will be selected, and in turn, the directors elect a slate of officers (managers), adopt by-laws, and arrange for the capitalization of the corporation. The by-laws can serve as an internal charter for authority of officers, timing and conduct of shareholder and board meetings, or other corporate details. A corporation, as an entity of the state of its formation, is an artificial person with "domicile" in that state, and if it intends to conduct business in other states, the corporation must be qualified as a "foreign corporation" in such state or states.

9.9 CORPORATION: OWNERSHIP, CONTROL, AND MANAGEMENT

Stockholders Ownership and Capitalization

While the stockholders are the owners of the corporation, they have a limited role in making business decisions and managing the affairs of the corporation. The stockholders exercise influence over broad policy and manage indirectly by selecting the directors.

Certain transactions of the corporation require stockholder approval by affirmative vote, such as (1) the sale of all the corporate assets, (2) the merger of the corporation with another corporation, (3) amendment of the corporation's articles of incorporation, and (4) voluntary dissolution of the corporation. There has been a recent trend in publicly held corporations for the shareholders to place propositions on the ballot at the annual stockholder meeting relating to matters of social policy.

Board of Directors

All corporate statutes provide that the corporation's affairs are to be managed by its board of directors. The board has power to enter contracts, borrow money, mortgage property, select officers, establish their compensation, and declare dividends.

The board meets at regular intervals and may have special meetings on call. Because the authority to manage the affairs of the corporation is vested in the directors as a board and does not reside in the directors as individuals, the directors can neither act nor bind the corporation as individual directors since corporate authority requires the deliberative decision based on combined judgment of the entire board. Since the board acts as a deliberative body, corporate law requires notice to each director of the time, place, and purpose of the meeting. Some modern statutes now permit board meetings by means of a telephone conference call after proper notice if all directors are connected on the call or have waived attendance.

Management of Corporation

While the board of directors retains ultimate control at the policy-making level, the board elects corporate officers who manage the corporation's day-to-day affairs. These officers are answerable to the board but are not directly responsible to the stockholders. The power of the officers is based on actual authority, express and implied, conferred on them by the board of directors; but, to protect innocent third parties dealing with the corporation, the courts also will hold the corporation liable for acts performed by officers and agents of the corporation based on their apparent authority. Further, the doctrine of ratification discussed earlier in this chapter is equally applicable.

9.10 CORPORATION: STOCKHOLDER IMMUNITY AND PIERCING THE CORPORATE VEIL

Stockholders are protected from personal liability for the corporation's debts (contract or tort) by the legal fiction that the corporation is an entity separate and distinct from its stockholders. However, in some instances, the court sets aside this legal fiction and concludes that the corporation is the alter ego of the stockholders and holds the stockholders personally liable for the debts of the corporation. The process of disregarding the corporate entity and holding its stockholders subject to unlimited liability is referred to as "piercing the corporate veil." There is no brightline test to determine when the corporate veil will be pierced, but, in most instances, it is due to a combination of factors that convince the court that the action is required to do equity and avoid fraud. Although a few courts have held otherwise, the weight of authority holds that the corporate entity will not be disregarded solely because the number of stockholders is only a few, even if the corporation is wholly owned by a single stockholder. The most common grounds for piercing the corporate veil are (1) failure to observe formalities of corporate form and (2) inadequate capitalization.

In situations where the corporation is operated as a mere conduit for the stockholder's personal affairs and if the stockholders seek to operate the corporation by *fiat* without observing the formalities of board of director's meetings, failing to keep minutes, and commingling personal affairs with corporate affairs, piercing the corporate veil is justified.

Turning to the question of inadequate capitalization, more difficult questions are raised. "Under capitalization" is a question of fact based on the type of business undertaken and the economic needs and risks of each business on a case-by-case basis. While some courts hold that a corporation is inadequately capitalized when its stockholders have not invested enough capital for the corporation's stock to meet its prospective business risks and economic needs, the application of this standard where the corporation has met the state's legal requirements including insurance coverage raises an interesting dilemma.

However, rarely is inadequate capitalization standing alone the cause of disregarding the corporate entity. Conduct such as where the stockholders have "milked" the corporation of capital by siphoning off profits and prevented it from accumulating capital to meet the corporation's expanding needs represents a more critical factor in causing the corporate veil to be pierced.

WALKOVSZKY v. CARLTON
276 N.Y.S.2d 585, 223 N.E.2d 6 (N.Y. 1966)

Fuld, Judge. [The complaint alleges that the plaintiff was injured when he was run down by a taxicab owned by defendant, Seon Cab Corporation, and negligently operated at the time by defendant Marchese. Defendant Carlton is alleged to be a stockholder of 10 corporations, including Seon, each of which has but two cabs registered in its name and carries the minimum amount ($10,000) of automobile liability insurance. It is alleged that these corporations are "operated as a single entity, unit and enterprise" with regard to financing, supplies, repairs, employees and garaging. Plaintiff contends that the defendant Carlton is personally liable because the multiple corporate structure constitutes an unlawful attempt "to defraud members of the public" who might be injured by the cabs.]

The law permits the incorporation of a business for the purpose of enabling its proprietors to escape personal liability but manifestly, the privilege is not without its limits. Broadly speaking the courts will disregard the corporate form, or, to use accepted terminology, "pierce the corporate veil", whenever necessary "to prevent fraud or to achieve equity."

We are guided, as Judge Cardozo noted by "general rules of agency." In other words, whenever anyone who uses control of the corporation to further his own rather than the corporation's business, he will be liable for the corporation's acts "upon the principle of *respondeat superior* applicable even where the agent is a natural person."

In the *Mangan case*, the plaintiff was injured as a result of the negligent operation of a cab owned and operated by one of the four corporations affiliated with the defendant Terminal. Although the defendant was not a stockholder of any of the operating companies, both the defendant and the operating companies were owned, for the most part, by the same parties. The defendant's name (Terminal) was conspicuously displayed on the sides of all the taxis used in the enterprise and, in point of fact, the defendant actually serviced, inspected, repaired and dispatched them. These facts were deemed to prove sufficient cause for piercing the corporate veil of the operating company, the nominal owner of the cab which injured the plaintiff, and holding the defendant liable. The operating companies were simple instrumentalities for carrying on the business of the defendant without imposing upon it financial and other liabilities incident to the actual ownership and operation of the cabs.

In the case before us, the plaintiff has explicitly alleged that none of the corporations "had a separate existence of their own" and, as indicated above, all are named as defendants. However, it is one thing to assert that a corporation is a fragment of a larger corporate combine which actually conducts the business. It is quite another to claim that the corporation is a "dummy" for its individual stockholders who are in reality carrying on the business in their personal capacities for purely personal rather than corporate ends. Either circumstance would justify treating the corporation as an agent and piercing the corporate veil to reach the principal but a different result would follow in each case. In the first, only a larger *corporate* entity would be held financially responsible . . . while, in the other, the stockholder should be personally liable. Either the stockholder is conducting the business in his individual capacity or he is not. If he is, he will be liable; if he is not, then it does not matter—insofar as his personal liability is concerned— that the enterprise is actually being carried on by a larger "enterprise entity." The individual defendant is charged with having "organized, managed, dominated and controlled" a fragmented corporate entity but there are no allegations that he was conducting business in his individual capacity. Had the taxicab fleet been owned by a single corporation, . . . the plaintiff would face formidable barriers in attempting to establish personal liability on the part of the corporation's stockholders. The fact that the fleet ownership has been deliberately split among many corporations does not ease the plaintiff's burden in that respect. The corporate form may not be disregarded merely because the assets of the corporation, together with the mandatory insurance coverage of the vehicle that struck the plaintiff, are insufficient to assure him the recovery sought. If the insurance coverage required by statute is inadequate for the protection of the public, the remedy lies not with the courts but with the legislature. It may well be sound policy to require that certain corporations must take liability insurance which will afford adequate compensation to their potential tort victims. However, the responsibility for imposing conditions on the privilege of incorporation has been committed by the Constitution to the Legislature.

In point of fact, the principle relied upon in the complaint to sustain the imposition of personal liability is not agency but fraud. Such a cause of action cannot withstand analysis. If it is not fraudulent for the owner-operator of a single cab corporation to take out only the minimum required liability insurance, the enterprise does not become illicit or fraudulent merely because it consists of many such corporations.

Keating, Judge (dissenting). From their inception these corporations were intentionally undercapitalized for the purpose of avoiding responsibility for acts which were bound to arise as a result of the operation of a large taxi fleet having cars out on the street 24 hours a day and engaged in public transportation. And during the course of the corporations' existence all income was continually drained out of the corporations for the same purpose.

The issue presented in this action is whether the policy of this State, which affords those desiring to engage in a business enterprise the privilege of limited liability through the use of the corporate device, is so strong that it will permit that privilege to continue no matter how much it is abused, no matter how irresponsibly the corporation is operated, no matter what the cost to the public. [Citing the California case of *Minton v. Cavaney*, Judge Keating relied on Judge Traynor's holding that liability should attach when the corporation is inadequately capitalized and the stockholders actively participate in the conduct of the corporate affairs.

MINTON v. CAVANEY
13 Cal. Rptr. 641, 364 P.2d 473 (1961)

Traynor, Justice. The Seminole Hot Springs Corporation, hereinafter referred to as Seminole, was duly incorporated in California on March 8, 1954. It conducted a public swimming pool that it leased from its owner. On June 24, 1954, plaintiffs' daughter drowned in the pool, and plaintiffs recovered a judgment for $10,000 against Seminole for her wrongful death. The judgment remains unsatisfied.

On January 30, 1957, plaintiffs brought the present action to hold defendant, Cavaney, personally liable for the judgment against Seminole. The trial court entered judgment for plaintiffs for $10,000. Defendant appeals.

Plaintiffs introduced evidence that Cavaney was a director and secretary and treasurer of Seminole and that on November 15, 1954, about five months after the drowning, Cavaney as secretary of Seminole and Edwin A. Kraft as president of Seminole applied for permission to issue three shares of Seminole stock, one share to be issued to Kraft, another to F. J. Wettrick and the third to Cavaney. The commissioner of corporations refused permission to issue these shares unless additional information was furnished. The application was then abandoned and no shares were ever issued. There was also evidence that for a time Seminole used Cavaney's office to keep records and to receive mail. Before his death Cavaney answered certain interrogatories. He was asked if Seminole "ever had any asset?" He stated that "insofar as my own personal knowledge and belief is concerned said corporation did not have any assets." Cavaney also stated in the return to an attempted execution that "[I]nsofar as I know, this corporation had no assets of any kind or character. The corporation was duly organized but never functioned as a corporation."

Defendant introduced evidence that Cavaney was an attorney at law, that he was approached by Kraft and Wettrick to form Seminole, and that he was the attorney for Seminole. Plaintiffs introduced Cavaney's answer to several interrogatories that he held the post of secretary and treasurer and director in a temporary capacity and as an accommodation to his client.

Defendant contends that the evidence does not support the court's determination that Cavaney is personally liable for Seminole's debts and that the "alter ego" doctrine is inapplicable because plaintiffs failed to show that there was "(1) . . . such unity of interest and ownership that the separate personalities of the corporation and the individual no longer exist and (2) that, if the acts are treated as those of the corporation alone, an inequitable result will follow."

The figurative terminology "alter ego" and "disregard of the corporate entity" is generally used to refer to the various situations that are an abuse of the corporate privilege. The equitable owners of a corporation, for example, are personally liable when they treat the assets of the corporation as their own and add or withdraw capital from the corporation at will . . . when they hold themselves out as being personally liable for the debts of the corporation . . . or when they provide inadequate capitalization and actively participate in the conduct of corporate affairs . . . .

In the instant case the evidence is undisputed that there was no attempt to provide adequate capitalization. Seminole never had any substantial assets. It leased the pool that it operated, and the lease was forfeited for failure to pay the rent. Its capital was "trifling compared with the business to be done and the risks of loss . . . ." The evidence is also undisputed that Cavaney was not only the secretary and treasurer of the corporation but was a director. The evidence that Cavaney was to receive one-third of the shares to be issued supports an inference that he was an equitable owner, and the evidence that for a time the records of the corporation were kept in Cavaney's office supports an inference that he actively participated in the conduct of the business. The trial court was not required to believe his statement that he was only a "temporary" director and officer "for accommodation." In any event it merely raised a conflict in the evidence that was resolved adversely to defendant. Moreover, . . . the Corporation Code provides that ". . . the business and affairs of every [corporation] shall be controlled by a board of not less than three directors." Defendant does not claim that Cavaney was a director with specialized duties. (See, 5 U.Chi.L.Rev. 668.) It is immaterial whether or not he accepted the office of director as an "accommodation" with the understanding that he would not exercise any of the duties of a director. A person may not in this manner divorce the responsibilities of a director from the statutory duties and powers of that office.

In this action to hold defendant personally liable upon the judgment against Seminole, plaintiffs did not allege or present evidence on the issue of Seminole's negligence or on the amount of damages sustained by plaintiffs. They relied solely on the judgment against Seminole. Defendant correctly contended that Cavaney or his estate cannot be held liable for the debts of Seminole without any opportunity to relitigate these issues. Cavaney was not a party to the action against the corporation, and the judgment in that action is therefore not binding upon him unless he controlled the litigation leading to the judgment.

[Reversed. Judgment for Plaintiff.]

Schauer, Justice (concurring and dissenting). I concur in the judgment of reversal on the ground stated in the majority opinion. I dissent from any implication that *mere professional activity by an attorney at law, as such,* in the organization of corporation, can constitute any basis for a finding that the corporation is the attorney's alter ego or that he is personally liable for its debts; whether based on contract or tort. In the process of developing an idea of a person or persons into an embryonic corporation and finally to full legal entity status with a permit issued, directors and officers elected, and assets in hand ready to begin business, there may often be delays. In such event a qualifying share of stock may stand in the name of the organizing attorney for substantial periods of time. In none of the activities indicated is the corporation actually engaging in business. And the lawyer who handles the task of determining and directing and participating in the steps appropriate to transforming the idea into a competent legal entity *ready to engage in business* is not an alter ego of the corporation. By his professional acts he has not been engaging in business in the name of the corporation; he has been merely practicing law.

WASHINGTON v. MECHANICS & TRADERS INS. CO.
50 P.2d 621 (Okla. 1935).

Action by Mechanics & Traders Insurance Company, plaintiff, against J. Wilson Washington, Virgil D. Carlisle, and Joe E., Edmundson, defendants. Joe E. Edmundson was the local policy-writing agent of the plaintiff. The other two defendants were sureties on his bond. Edmundson delivered to Sallie Young a policy of insurance upon her property which the plaintiff instructed him to cancel. This he neglected to do. The insured property was destroyed by fire. Plaintiff was compelled to pay and this action followed.

Per Curiam: The undisputed testimony in the record shows that the plaintiff instructed the said Edmondson [in October] to cancel the policy issued to Sallie Young and said instruction was repeated [three times in November]; yet, in view of these instructions, the agent, Edmondson, failed, neglected, or refused to cancel said policy. The further fact was undisputed that the house of the assured was destroyed by fire and that the husband of the assured, John Young, was in the agent's office the day before the fire, and that the cancellation of the policy was discussed and the agent, Edmondson, at that time did not cancel the policy when he had sufficient funds on hand to return the premiums, but instead of this [Edmundson] continued the policy in effect in direct conflict with his instructions. [Edmundson also] refused to talk the matter over with the general agent. Kline, of the insurance company, came to [Edmundson's] office the [day before] the property was destroyed and the plaintiff company had to pay the loss.

Where an agent, whose powers extend to the cancellation of policies, is directed by the company to cancel a policy, and he neglects to do so within a reasonable time, and in the meantime there has been a loss, he is liable to the company for the amount which the latter is compelled to pay on such loss, unless he can show some valid reason for his failure to follow the company's direction. His delay or failure to cancel the policy will not be excused by the fact that he believed that the company was mistaken as to the safety or danger of the risk, or as to the wisdom of retaining it, or by the fact that he gave notice of the cancellation to the broker who negotiated the insurance and directed him to cancel it.

The neglect and failure of the agent, Edmondson, to carry out the instructions of plaintiff company, to cancel the Sallie Young policy, for more than five weeks, was not the exercise of good faith and reasonable diligence, as to which reasonable men could differ, and the fact that the plaintiff company had the same authority to cancel the policy would not excuse or justify the agent, Edmondson, in failing or refusing to follow its instructions. [Decision for Plaintiff.]

NATIONAL BISCUIT COMPANY v. STROUT
106 S.E. 2d 692 (N. C. 1959)

Parker, Justice. C. N. Stroud and Earl Freeman entered into a general partnership to sell groceries under the firm name of Stroud's Food Center. There is nothing in the agreed statement of facts to indicate or suggest that Freeman's power and authority as a general partner were in any way restricted or limited by the articles of partnership. Certainly, the purchase and sale of bread were ordinary and legitimate businesses of Stroud's Food Center during its continuance as a going concern.

Several months prior to February 1956, Stroud advised plaintiff that he personally would not be responsible for any additional bread sold by plaintiff to Stroud's Food Center. After such notice to plaintiff, it from 6 February 1956 to 25 February 1956, at the request of Freeman, sold and delivered bread in the amount of $171.04 to Stroud's Food Center.

In [an earlier case], this Court said: "A and B are general partners to do some given business; the partnership is, by operation of law, a power to each to bind the partnership in any manner legitimate to the business. If one partner goes to a third person to buy an article on time for the partnership, the other partner cannot prevent it by writing to the third person not to sell to him on time; or, if one party attempts to buy for cash, the other has no right to require that it shall be on time. And what is true in regard to buying is true in regard to selling. What either partner does with a third person is binding on the partnership. It is otherwise where the partnership is not general, but is upon special terms, as that purchases and sales must be with and for cash. There the power to each is special, in regard to all dealings with third persons at least who have notice of the terms." There is contrary authority. 68 C.J.S. Partnership Section 143. pp. 578-579. However, this text of C.J.S. does not mention the effect of the Provisions of the Uniform Partnership Act.

The General Assembly of North Carolina . . . enacted a Uniform Partnership Act, and [The court . . . reads UPA Section 15 as] "Every partner is an agent of the partnership for the purposes of its business . . . unless the partner so acting had no authority to act for the partnership in the particular matter and the person with whom [the partner] is dealing has knowledge of the fact that [the partner] has no such authority."

Freeman, as a general partner with Stroud, with no restrictions on his authority to act within the scope of the partnership business so far as the agreed statement of facts showed, had under the Uniform Partnership Act "equal rights in the management and conduct of the partnership business." Under [the UPA], Stroud, his copartner, could not restrict the power and authority of Freeman to buy bread for the partnership as a going concern, for such a purchase was an "ordinary matter connected with the partnership business," for the purpose of its business and within its scope, because in the very nature of things Stroud was not, and could not be, a majority of the partners. Therefore, Freeman's purchases of bread from plaintiff for Stroud's Food Center as a going concern bound the partnership and his copartner Stroud.

In Crane on Partnership, 2d Ed., p. 277, it is said: "In cases of an even division of the partners as to whether or not an act within the scope of the business should be done, of which disagreement a third person has knowledge, it seems that logically no restriction can be placed upon the power to act. The partnership being a going concern, activities within the scope of the business should not be limited, save by the expressed will of the majority deciding a disputed question; half of the members are not a majority."

At the close of business on 25 February 1956 Stroud and Freeman by agreement dissolved the partnership. By their dissolution agreement all of the partnership assets, including cash on hand, bank deposits and all accounts receivable, with a few exceptions, were assigned to Stroud, who bound himself by such written agreement to liquidate the firm's assets and discharge its liabilities. It would seem a fair inference from the agreed statement of facts that the partnership got the benefits of the bread sold and delivered by plaintiff to Stroud's Food Center, at Freeman's request, from 6 February to 25 February 1956. But whether it did or not, Freeman's acts, as stated above, bound the partnership and Stroud.

The judgment of the court below is: Affirmed for plaintiff.

DISCUSSION QUESTIONS

1. Discuss the pros and cons of doing business in the form of (1) a corporation and (2) a partnership.

2. Under what circumstances may the stockholders be held personally liable for the debts or torts of the corporation?

3. Describe the difference between an agency/employee relationship and the "independent contractor" relationship. What difference does it make in terms of liability claims against the principal or the employer?

4. Describe the meaning of "implied authority" and "apparent authority" with respect to an agent, a partner, and a corporate officer.

5. On what basis and with what success could the stockholders of a large publicly held corporation complain about a substantial undesignated and unrestricted gift to a major private university?

6. What are the legal and ethical questions in a situation involving an architect, who is on retainer with an airline to design airline facilities for use in various airports, when the architect accepts an assignment from a major

city airport authority to design a new passenger and baggage facility for the airport authority?

7. X is in fact the sole proprietor of an engineering firm, Y. Z, a professional engineer, is employed by firm Y and is engaged in doing design work. One day, X and Z went by the bank that handled the account of firm Y. X introduced Z as "my partner" to the bank president. The following month, Z negotiated a loan from the bank on behalf of firm Y, which loan was executed by X. Subsequently, firm Y defaults on its obligations to the bank. Discuss the bank's claims to hold X, firm Y, and Z responsible.

10

CONSTRUCTION CONTRACTS

Objectives

Expenditures for new construction in the United States including residential home construction annually exceed $350 billion. The construction industry has a very important economic impact on a number of related industries. In turn, the national and regional economies are directly affected by the health of the construction industry since the construction industry has a significant effect on suppliers of materials, equipment, and labor.

A construction project is a complex undertaking, and the construction contract serves as the centerpiece of the entire project. While construction contracts have characteristics common to other contracts, construction contracts have their own unique features. The purpose of the construction contract is to implement the objectives of the principal parties including the owner, tenant-occupier, contractor, design professional, and lending institution and to govern their rights, duties, and liabilities. While the owner's desire is sometimes articulated as "a good job, on time and within budget," it is otherwise more formally expressed as an expectation that the project be completed on schedule and also perform and serve the anticipated purpose. The contractor seeks to obtain the expected profit within the agreed compensation. The contractor's role is frequently circumscribed by the fact that the contractor is not selected until plans and specifications are complete, limiting input on contractual considerations. Obviously, there are many players, including architects, design professionals, consultants (e.g., environmental, structural, or safety experts), lenders, and suppliers, all serving important roles in the successful completion of the project.

While construction contracts usually contain many clauses frequently referred to as "boiler plate" or "standard clauses," it is important that their meaning be clearly understood and coordinated with the individualized parts of the contract. The parties benefit if the contract is properly drawn, the specifications are clear and complete, and the drawings are properly finished in sufficient detail so that the contractor has no questions concerning what is expected by the owner.

Parties

Owners do not fall into a single common mold. They may be (1) private or public, (2) owner-occupier or owner-landlord, or (3) experienced professional owner or novice inexperienced owner (or any shading between those extremes). Where the owner is the federal, state, or city government or an agency of a governmental unit, the project is usually referred to as "public work."

The private/public dichotomy is important since the private owner, as contrasted with the public owner, has a great deal more latitude in selecting design professionals and contractors either by competitive bidding or by negotiation. Private owners are motivated by economic factors such as financing, return on investment, profitability, and market risk. On the other hand, the public owner who uses public funds, commonly is restricted by statutes and regulations in selecting contractors based on the lowest bid. Public owners' objectives, which are performance and compliance oriented, are tempered by cost limitations dictated by appropriated public funds.

It is also obvious that the contractor should take a decidedly different approach to contract documents and construction work when the project involves a building for a major manufacturing concern as contrasted with a firm of attorneys or doctors who for the first time have decided to build and own their own office building. In the latter case, a greater burden tacitly falls on the contractor where the "other party" lacks construction sophistication.

From the owner's perspective, the traditional concept of the contractor is the organization and completion of the project in accordance with the plans and specifications. Usually, the general contractor is a single entity, whether a sole entrepreneur, partnership, or corporation. The general contractor (sometimes referred to as the prime contractor) serves as manager of both that portion of the work that is accomplished by the prime contractor's own forces and the work performed by subcontractors (sometimes referred to as speciality contractors such as plumbing, electrical, iron work). While the design work may be performed by either the owner or an independent professional engaged by the owner, it also may be performed by the contractor. Under any circumstances, it is useful to have the contractor involved during the design phase because of the contractor's knowledge of construction techniques, materials, and labor matters.

The role and responsibility of the design professional or the architect with respect to performance during the construction period is frequently limited by contractual clauses that tend to be exculpatory clauses. In *Kreiger v. J. E. Greiner Co.*, 382 A.2d 1069 (Md. 1978), where an employee was injured when a steel column collapsed, the court interpreted provisions of the General Conditions, reading "[that the defendant consulting engineer] expressly agreed to see that the method of construction work conform to all federal, state, and local laws and regulations" did not mean that the consulting engineer had a duty "to see that the contractors obey all laws in connection with the work." In denying recovery to the injured employee, the court concluded that there was no provision in the contract imposing a duty on the defendant consulting engineer to supervise the method of construction. In contrast, *Caldwell v. Bechtel, Inc.*, 631 F.2d 989 (D.C. Cir. 1980), suggests that "unlike contract duties, which are imposed by agreement of the parties to a contract, a duty of care under tort law is based primarily upon social policy." The court referred to "societal expectations" as the conceptual basis which "comprises and defines the limits of Bechtel's duties." In that case, the court held Bechtel, who served as a consultant performing "safety engineering" services, liable for silicosis suffered by contractor's employee, resulting from silica dust when the ventilation in the Metro tunnels was inadequate. The decision emphasized that "Bechtel's superior skills and position [resulted] in Bechtel's ability to foresee the harm that might be expected to befall plaintiff [creating] a duty in Bechtel to take reasonable steps to prevent harm to [plaintiff] from the hazardous conditions of the subway tunnels."

10.2 COMPENSATION ARRANGEMENTS

Fixed-Price Contract

Under the fixed-price contract, the contractor agrees to do the work called for in the construction contract for a stated number of dollars by submitting a single lump sum price for the completed project. It is important to note that, absent agreement to the contrary, the contractor agrees to perform the required work for an agreed dollar amount regardless of any delays, problems encountered, or unanticipated expenses. This style of contract may be successfully used for construction of projects of the type and size in which the contractor has considerable experience. In administering this type of contract, the allocation of progress payments is a matter of concern to avoid "front loading," by which early work is overpriced to be offset by underpricing later work to the cash flow advantage of the contractor at the expense of the owner. Occasionally, the fixed-price contract will contain a "shared-savings" clause under which the contractor and owner will divide, on an agreed percentage or other basis, any savings that the contractor is able to achieve below the contract price.

Unit-Price Contract

Under the unit-price contract which is popular in public works, the contractor's compensation is based on the quantities of specified materials used in the project. The parties may estimate the contract price based on their appraisal of the number of units that will be used. However, the contractor's final compensation will be based on the quantity of units required and placed in the finished work (e.g., cubic yards of concrete, cubic yards of excavation, or so many tons of pavement). From the owner's viewpoint, the disadvantage of this type of contract is that the final contract price cannot be ascertained until the project is completed. On the other hand, it permits the owner to avoid preparing numerous preliminary contract drawings. Supplemental drawings prepared as the work progresses will suffice, with timing dictated only by the necessity of permitting the contractor to secure materials.

Cost-Plus Contract

Under the cost-plus contract, the contractor is paid the actual cost of the work plus an additional sum covering overhead and profit. This type of contract requires detailed record keeping of costs, which frequently includes materials, labor, rental of equipment and property, transportation, and numerous other items with the possible exception of the contractor's supervisory staff. Under this type of contract, record keeping shifts from work completed to record keeping of costs for doing work resulting in a vast and detailed set of records. The "plus" may be a flat fee or calculated as a percentage of cost. Because of the open-ended nature of this arrangement, owners frequently demand a maximum or upset price. From the contractor's standpoint, this type of contract is useful when the project involves uncertain techniques or conditions such as new materials or designs. This type of contract is considered an excellent method of getting work done rapidly and with a respected experienced contractor who has a reputation for working efficiently.

In cost-plus contracts, where the owner has an upset price, the agreement may require that the contractor notify the owner when it appears that the cost will exceed a given figure. In *Jones v. J. H. Hiser Construction Co., Inc.* 484 A.2d 302 (Md. App. 1984), involving a cost-plus fixed-fee contract, the court held that there was an implied obligation on the contractor "to know of, keep track of, and advise the owner that actual costs were substantially exceeding estimates" and failure to do so barred recovery on the "cost overruns."

10.3 CONTRACT DOCUMENTS

Feasibility Studies

A major construction project has several identifiable phases including (1) feasibility study, (2) preliminary/final design, (3) bidding or negotiation, (4) construction, and (5) close out. In any well-managed and planned construction project, except emergency situations, the first step is the preparation of the architectural concept and preliminary engineering as well as a marketing and economic investigation. This process undertakes to match the final structure to the economic needs of the project. In addition, the feasibility studies help assess specific construction design and construction problems, provide analysis of the owner's needs, study various locations and alternatives, identify government requirements, and consider general socioeconomic and environmental effects.

In these agreements, the engineer must be alert to prevent the inclusion of language that would broaden the traditional duty to the owner or third parties. It is especially important to avoid language that could be interpreted as a guaranty, since professional insurance does not ordinarily cover guarantees, and any expansion of duties compounds the problem.

Components of the Construction Documents

The contract documents generally are composed of the agreement between the contractor and the owner, the bid, the drawings, the specifications, the bonds, and other relevant instruments, such as general and special terms and conditions. As a matter of caution, legal counsel may seek to provide a priority of the documents in the event there is a conflict among them. Dating back as early as the 1940s, various engineering societies, including the American Society of Civil Engineers (ASCE), Associated General Contractors of America (AGCA), and the National Society of Professional Engineers (NSPE), have studied, drafted, and sponsored standard contract forms. In developing its publications, the Engineers Joint Contract Document Committee (EJCDC) receives input from design professionals, owners, contractors, and attorneys and takes note of related legislation and court decisions. In addition to these documents endorsed by the ASCE, the American Institute of Architects (AIA) prepares model construction documents. However, while the use of standardized documents has increased during the 1980s, there are no standard documents that are in general use throughout the United States.

An important aspect of the main contract is the "scope of work" provision that defines the work to be done by the contractor, and the more detailed and specific those responsibilities are set forth, the less chance for misunderstanding and dispute. Every effort should be made to avoid open-ended language such as "all alternatives will be studied." Customarily, the contractor will agree to do all of the defined work including the furnishing of labor, materials, and equipment necessary for completion of the project. These agreements are interpreted to include not only the work specified in the plans and specifications but also the work that is "reasonably implied" by all of the construction documents.

Change orders. Projects invariably change as construction is in progress. Even on a simple family dwelling, and more so as the size of the project increases, the owner will think of additions or want to revise existing plans. If the contract documents do not provide a change order procedure, the parties will have to devise an ad hoc procedure. In some cases, the change may be initiated by the contractor with regard to contract price or time of performance. Customarily, the design professional will review change orders and then discuss them with the owner recommending acceptance or rejection. If the change order is accepted by the parties, it should be signed by properly authorized representatives to become part of the contract documents. Since there is often disagreement as to value or cost when changes occur, a technique should be established to reach a prompt and objective resolution of the dispute. Frequently at issue is the authority of the individual who specifies additional work or performance in a different manner and, if the request occurs at the job site, the procedure may not be in compliance with the formality of the contract documents. These questions should ordinarily be resolved on the basis of the individual's authority to bind the principal under traditional agency doctrines. The agent's authority is discussed in chapter 9.

Owner-requested change orders can be abused either by a major change in the entire project or the accumulated number of "nibbling" small change orders. Since most contractor's have several jobs underway at the same time, with coordinated flow sheets for the various jobs, a major extension or numerous small but continuous revisions seriously interfere with the planning process of the contractor. Such a situation raises questions as to whether the contractor may treat that state of affairs as a material breach by the owner or whether the work should be completed and a claim made for the reasonable value of services without regard to the contract price.

Design drawings. Broadly speaking, the design drawings show the design concept, size, and scope of job, number and size of materials or items, and how they are assembled in the completed project. They serve the purpose of giving the contractor information that can be communicated more effectively by pictures than by words. This is especially true with respect to shapes as well as the location and relationship of various parts within the project. Compliance with design drawings usually relieves the contractor of liability if the project is unsuccessful or does not meet the owner's expectations.

The function of the drawings is to make the subject as clear and definite as the draftsperson is able to do. On the other hand, it is difficult to do drawings so extensive as to show every detail of the project. However, it is essential that the drawings be in sufficient detail and accuracy to enable an experienced contractor to understand the project. The quality and accuracy of the drawings are useful and important as the contractor considers the scope of the work.

Specifications. Since specifications generally contain information on materials, performance necessary, and quality requirements, their preparation require the unique skill of identifying particular items in written form. Specifications should be cast in clear concise language avoiding legalese, generalities, or ambiguities.

At the outset of the specifications, it is useful to describe the work to be covered such as "removal and replacement of the piling which supports the superstructure of the existing bridge."

In describing standard products, it is helpful to give the trade name, catalog number, and any suitable specification of size, capacity, etc., such as,

Tubs Standard P325 5P and Q1001 with 200 gallon capacity. This specification covers the complete installation as manufactured by Acme Products, or equal as approved by the Architect.

In some cases, the specifications may describe the conditions or service that the proposed material will be required to meet or perform, such as,

Insulation which will resist a certain transfer of heat and be rigid, fire resistant, moisture proof and non decaying.

In addition to the specifications for materials, the specifications may describe the workmanship or the manner in which work is to be performed. For example,

Concrete shall be poured only during the presence of the inspector and by methods approved by the Owner's engineer or construction manager. All concrete shall be placed *in the dry*. Should water accumulate in any place where concrete is to be poured, the Contractor shall provide and operate sufficient pumps and do whatever else is necessary to remove the water in an approved manner. Water (other than that used for curing) shall be prevented from coming into contact with concrete while it is setting.

"Or-equal" provisions. The specifications may designate the products or equipment by manufacturer name and type. Requests for substitution are commonplace, and some contracts establish procedures by which the contractor may obtain approval to use alternate products or material. In other cases, the contract will contain a blanket "or-equal" clause, allowing the contractor to substitute material or equipment if of equal quality. Generally, if the substitute is obtained at a lower price, or the substitute is not equal but acceptable, an appropriate adjustment in cost should be made for the owner.

Unless owner's consent is obtained, the substitution raises a legal question of whether the use of the substitute product or material shifts the design risk to the contractor.

10.4   Concepts of Construction or Project Manager

The traditional system of construction separated the design process from the construction process. In that system, the design process was usually performed by an independent design professional; in turn, the construction process was performed by contractors usually involving a prime or general contractor and frequently speciality subcontractors. This separation of design and construction had the weakness of depriving the owner of some specialized skills of the contractor with respect to knowledge of labor and material markets as well as the advantages and disadvantages of various construction techniques.

Since construction costs are dramatically affected by costs of financing and delays in commencing facility use, owners have intensified their efforts to compress the construction time required by traditional methods. As a result, a new trend in construction responsibility began to change the traditional three-legged stool relationship involving three separate and distinct parties: owner, design professional (including architectural planning), and general contractor. In the traditional mode, the design professional developed the plans and specifications at the direction of the owner, who provided a broad brush expression of the needed facility and the financial backing. In turn, the plans and specifications were submitted to the general contractor for construction. In this traditional division of labor context, architect/designers were blamed for their lack of sensitivity toward costs and their lack of knowledge concerning labor and material markets. At the same time, contractors were blamed for their lack of skills or efforts in using new construction techniques and materials.

While the lines of jurisdiction were sometimes a bit fuzzy or overlapping under the traditional concept, the lines of authority were generally fairly clear. On the other hand, with the advent of the so called "construction manager" (CM), some of the traditional relationships shifted, raising such questions as the licensing requirements of the CM, matters involving conflict of interest when the general contractor serves as CM, and the scope of the CM's liability.

Definition of Construction Manager

The federal government in late 1980 through the Federal Procurement Regulations Directorate, Office of Acquisition Policy, issued a proposal to define the scope and meaning of "construction management." The definition subpart reads

(a) "Construction Management concern" means a concern that is an experienced, multidisciplined organization or joint venture qualified to accomplish selected construction management (professional) services. The ability to perform services is acquired through extensive experience in all phases of the construction process."

(b) "Construction management services" . . . encompasses a wide range of services relating to the management of a project during the predesign, design, and construction phases. The types of services include development of project strategy . . . scheduling . . . observation to insure that workmanship and materials comply with the plans and specifications . . . .

While the term construction manager seems quite straight forward, some of the standard documents have not embraced openly that term but opted for the term "project manager" (PM). For example, the "Standard Form of Agreement between Owner and Project Manager for Professional Services" (Document 1910-15 of the Engineers' Joint Contract Documents Committee) reads,

> The attached Agreement contemplates the combination of the design and management function from the start of the relationship with the Owner and the continuation thereof until completion of the construction. While carefully coordinating all aspects of his services with the Owner, the Project Manager will take over from the Owner many administrative and coordinating functions. He will, in effect, take charge of the Project from beginning to end in order to provide special expertise and relieve the Owner of duties and responsibilities which the Owner is neither qualified to undertake nor for which he will have a continuing need. However, under the attached Agreement *the Project Manager will not perform any of the functions of the contractors.* He may assist in purchasing on the Owner's behalf, coordinating the work of separate contractors and assist in expediting various aspects of the work; *but neither in his capacity as a professional engineer nor as a construction manager will he be involved in or assume responsibility for the means, methods, techniques, sequences, or procedures of construction or the safety precautions or programs incident thereto* [emphasis added].

Requirement of State License

Since the statutes of many states provide that the practice of engineering requires a valid state license, the activities of the CM raises questions as to the necessity of obtaining an appropriate license for professional practice as a construction manager. In 1974, the attorney general of California expressed the opinion that

> [I]t is unlawful for any person to engage in the business of contracting without a license . . . [noting . . . that a contractor is defined as] . . . any person, who undertakes to . . . or does himself or by or through others, construct, alter, repair, add to, subtract from, improve . . . any building, highway, road, parking facility . . . or other structure, project, development or improvement, or to do any part thereof.

Nevertheless, the attorney general concluded that the licensing requirement did not apply to the CM because "[a] construction manager does not bind himself to construct a building." *57 Op. Att'y. Gen. Cal. 421, 422 (1974).* In a similar vein, the Arkansas Supreme Court held that the CM is not required to hold a valid license as a contractor where there was a general contractor involved in the construction. The court reasoned,

> [I]n this particular case *there was a general contractor who had the supervisory authority,* and because of that fact, the services that the defendant undertook to perform, and did perform, under its contract as construction manager . . . do not fall with the definition of *contractor,* as that term if defined in [the state licensing statute], and the defendant was not, therefore, in violation of the Arkansas Contractors Law and not required to have a contractor's license in order to perform such services [emphasis added].

However, at the prodding of the state contractors group, the Arkansas legislature amended the licensing statute adding to its coverage "to manage the construction." Ark. Stat. Ann. Section 71-701 (1979). Additional aspects concerning the licensing requirement of construction or project managers as well as the potential for conflict of interest questions are discussed in a scholarly article by Milton F. Lunch, "New Construction Methods and New Roles for Engineers." 46 *Law and Contemporary Problems* 83 (1983).

"No Damages for Delay" Clauses

While the CM/PM is normally contractually obligated to schedule and coordinate the project work, there has been a contractual effort to absolve the contractor/CM/PM/owner from responsibility for the breach of those duties. Typically, "no damages for delay" clauses are drafted to restrict the specialty contractor's remedies for any delays to noncompensable time extensions regardless of the cause, duration, or cost of the delays. A typical clause reads,

> In any event [when] the Subcontractor is delayed in the commencement, prosecution or completion of its work by any act, omission, neglect or default of the Owner, Prime Contractor or anyone else employed by them, or of any other subcontractor or third party on the Project, or by any damage caused by fire or other casualty or by any extraordinary conditions arising out of war or governmental actions, or by any other cause all beyond the control and without the fault or neglect of Subcontractor, then Subcontractor shall be entitled to a noncompensable time extension only.

However, given a significant purpose of the use of professional construction management to shorten the time of project completion, the courts carefully scrutinize the "no damage delay" clause. In interpreting the clause, the court, in refusing to give effect to the clause, may consider that the CM's conduct amounted to "active interference" or may regard the delay "as beyond the contemplation of the parties" or "of unreasonable duration."

10.5   FAST TRACK/DESIGN BUILD PHASED CONSTRUCTION

In seeking new techniques to improve the economics and quality of construction projects, the construction industry initiated not only use of professional construction management but also the fast track/design build phased construction methods to reduce the time between the date of decision to build and the date of project completion.

The so-called fast/track technique essentially uses phased construction under which the owner may conceive of an idea and proceed with the project without delaying long enough to prepare a fixed set of plans and specifications for the entire project. Essentially, after the owner outlines for the architect the criteria for the project, the architect prepares, in varying degrees of detail, basic design documents, schematics, and preliminary drawings for plumbing, electrical, and structural drawings. In the fast track/design build mode, the contractor, based on these incomplete plans and specifications, attempts to estimate the cost of the construction. For example, in the traditional system, the structural members are designed from the top down, with the foundations being designed last. But, since construction begins with the foundation, the fast/track schedule requires that the foundations and lower columns be designed and built first before the upper levels are designed.

The advantage of this approach is that the design and construction phases overlap, which greatly compresses the total time from conception of the idea to project completion. The contractor can start the initial construction work prior to the completion and refinement of the total project design. Reducing construction time will provide savings for the owner by speeding the time when the project will begin production of rental income or the manufacture of goods. However, design professionals must take appropriate precautions to provide "quality" design.

The fast tract technique is suitable only for a sophisticated, experienced, and knowledgeable owner. This technique requires the utmost cooperation among the architect, general contractor, and the owner.

10.6 DEFENSES AGAINST OWNER CLAIMS

Since the standard AIA owner-architect agreement contains no "warranty" clauses imposing contractual guarantees of performance upon the architect, the liability of the architect is usually measured by the "standard of professional care." *Swett v. Gribaldo, Jones & Associates*, 115 Cal. Rptr. 99 (1974).

The AIA warranty does not impose responsibility on a contractor for defects resulting solely from design errors when the contractor performed in strict accordance with the plans and specifications furnished by the architect unless the contractor assumed greater responsibility. *Teufel v. Wiener*, 411 P.2d 151 (Wash. 1966). The principal difficulty results from the use of performance guarantees, which must be construed in connection with the "non warranty clauses" to determine the parties intentions and give weight to policy considerations. *City of Mac Pherson v. Stucker*, 256 P. 963 (Kan. 1927).

CALDWELL v. BECHTEL, INC.
631 F.2d 989 (Cir. D.C. 1980)

MacKinnon, Circuit Judge: [This case involves] a claim for damages by a worker who allegedly contracted silicosis while he was mucking in a tunnel under construction as part of the metropolitan subway system. The basic issue is whether a consultant engineering firm owed the worker a duty to protect him against unreasonable risk of harm.

The Shea-S&M-Ball joint venture (hereinafter Shea) entered into a contract with the Washington Metropolitan Area Transit Authority (WMATA) in the construction of tunnels for the Washington Metro Subway system and appellant Clem Caldwell was employed by Shea as a heavy equipment operator. Appellee Bechtel, Inc. (hereinafter Bechtel) was also under contract with WMATA, as a consultant to provide, inter alia, "safety engineering services" with respect to work to be done by various contractors pursuant to their contracts with WMATA. Among the duties that Bechtel undertook to perform for and on behalf of WMATA under the contract between the parties was the function of overseeing the enforcement of safety provisions in relevant safety codes, and inspecting job sites for violations. This appeal attacks the district court's grant of defendant Bechtel's motion for summary judgment . . . .

The essence of Caldwell's complaint is that Bechtel "had the function, duty and responsibility, as consultant to Metro (WMATA) to provide, inter alia, overall direction and supervision of safety measures and regulations in effect, or needed during the course of construction . . ." and that Bechtel was aware or should have been aware of the danger posed by high levels of silica dust and inadequate ventilation in the Metro tunnels, but failed to take the steps it was duty bound to take to rectify the situation. This failure, Caldwell alleges, was not only in violation of Bechtel's contract with WMATA and applicable safety codes, but more importantly constituted gross negligence toward Caldwell who was working in the tunnels.

While we must accept Caldwell's allegations of fact, we will closely examine his version of the law, since it stands in contradiction to the district court's conclusion "that the WMATA-Bechtel Contract created no duty owed plaintiff by defendant the breach of which would give rise to plaintiff's action for negligence." The issue in this case, then, is whether the contractual authority vested in Bechtel with respect to job site safety regulations created a special relationship between Bechtel and Caldwell under which Bechtel owed a duty to Caldwell to take reasonable steps to protect him from the foreseeable risk to his health posed by the dust laden Metro tunnels. We find that under applicable tort law principles Bechtel was indeed duty-bound. Accordingly, we reverse the decision of the district court.

In our view, the analysis of both Bechtel and the district court is overly reliant upon contract theory to the point of losing focus of the nature of the claim made here, which asserts negligence, rather than breach of contract. It has been many years since courts required privity of contract between the plaintiff and defendant before assessing tort liability, yet Bechtel would return us to that distant day. The duties that Bechtel undertook in its contract with WMATA are relevant to this case, not because they illustrate Bechtel's point that a contractual duty was owed to WMATA, but because by assuming a contractual duty to WMATA, Bechtel placed itself in the position of assuming a duty to appellant in tort. The particular circumstances of this case, including the Bechtel-WMATA contract, Bechtel's superior skills and position, resultant ability to foresee the harm that might reasonably be expected to befall appellant, created a duty in Bechtel to take reasonable steps to prevent harm to appellant from the hazardous conditions of the subway tunnels.

[In order to show that Bechtel] . . . owes no duty of reasonable care to protect [plaintiff] appellant's safety, Bechtel argues that by its contract with WMATA it assumed duties only to WMATA. Appellant [plaintiff] has not brought this action, however, for breach of contract but rather seeks damages for an asserted breach of the duty of reasonable care. Unlike contract duties, which are imposed by agreement of the parties to a contract, a duty of due care under tort is upon social policy. The law imposes upon individuals certain expectations of conduct, such as the expectancy that their actions will not cause foreseeable injury to another. These societal expectations, as formed through the common law, comprise the concept of duty.

Society's expectations, and the concomitant duties imposed, vary in response to the activity engaged in by the defendant. If defendant is driving a car, he will be held to exercise the degree of care normally exercised by a reasonable person in like circumstances. Or if defendant is engaged in the practice of his profession, he will be held to exercise a degree of care consistent with his superior knowledge and skill. Hence, when defendant Bechtel engaged in consulting engineering services, the company was required to observe a standard of care ordinarily adhered to by one providing such services, possessing such skill and expertise.

A secondary but equally important principle involved in a determination of duty is to whom the duty is owed. The answer to this question is usually framed in terms of the foreseeable plaintiff, in other words, one who might foreseeably be injured by defendant's conduct. This secondary principle also serves to distinguish tort law from contract law. While in contract law, only one to whom the contract specifies that the duty be rendered will have a cause of action for its breach, in tort law, society, not the contract, specifies to whom the duty is owed, and this has traditionally been the foreseeable plaintiff.

It is important to keep these differences between contract and tort duties in mind when examining whether Bechtel's undertaking of contractual duties to WMATA created a duty of reasonable care toward Caldwell.

Analyzing the common law, Prosser noted that courts have found a duty to act for the protection of another when certain relationships exist, such as carrier-passenger, innkeeper-guest, shipper-seaman, employer-employee, shopkeeper-visitor, host-social guest, jailor-prisoner, and school-pupil. These holdings suggest that courts have been eroding the general rule that there is no duty to act to help another in distress, by creating exceptions based upon a relationship between the actors.

The first component of the duty that Bechtel owed to appellant [plaintiff] takes as its point of departure Bechtel's contractual duties to WMATA. Under the contract Bechtel was to provide:

safety engineering services as required to ensure compliance with the provisions of the Metro Construction Safety Manual, the Metro Coordinated Safety Program and Reporting procedures and other applicable codes, and the actual obligation of the Authority's contractors, and [Bechtel] shall direct the contractor to correct any unsatisfactory condition which may be detected.

Bechtel was also "to develop and ensure a uniform system of safety and accident prevention procedures including reporting requirements."

Several duties are encompassed in these contractual terms. First, Bechtel was charged to ensure compliance with the Safety Manual, which included among its admonitions that "[n]o man shall be required to work in an unsafe place." Second, Bechtel's contract directed Bechtel's Resident Engineer to "report on unsafe working conditions" and to receive and monitor copies of the contractor's daily safety inspection reports and atmospheric logs, and gave the Resident Engineer authority to order work stopped "if unsafe conditions exist until such time as the condition is corrected."

Third, Bechtel was to ensure, as defined, that the other contractors on the job obeyed safety requirements and fulfilled contractual obligations to WMATA. One such duty owed by Shea to WMATA was to ensure "that the methods of performing work do not involve undue danger to the personnel employed thereon . . . ." And the responsibility fell to Bechtel to "direct [as defined] the contractor to correct any unsatisfactory condition which may be detected."

While the Bechtel-WMATA contract requires only that Bechtel use its "best efforts to persuade (Shea) . . . to comply" with safety regulations and thus Bechtel would not be absolutely liable to WMATA in the event of a safety violation, we are not only concerned with assessing Bechtel's duties to WMATA. Rather, the significance of the Bechtel-WMATA contract is that once Bechtel undertook responsibility for overseeing safety compliance, it assumed a duty of reasonable care in carrying out such duties that extended to the workers on the site.

In the case that forged the limits of the common law concept of duty, *Palsgraf v. Long Island R. Co.*, 162 N.E. 99 (N.Y. 1928), Chief Judge Cardozo declared that "the orbit of the danger as disclosed to the eye of reasonable vigilance would be the orbit of duty." [Cardozo added:] . . . "[t]he risk reasonably to be perceived defines the duty to be obeyed, and risk imports relation; it is risk to another or to others within the range of apprehension."

With these time-honored principles in mind, we have no difficulty in concluding that appellant was "within the orbit of danger" and hence within "the orbit of duty" owed by Bechtel. The primary workplace of course was the system of Metro tunnels under construction, the source of calculable danger to those employed there. If unsafe conditions were allowed to exist in the tunnels, injury to those engaged in the construction process was foreseeable. This was especially true since the tunnels were shown by atmospheric testing to contain a dangerous level of silica dust. Hence, Bechtel should be held to a duty to perceive and take reasonable steps to rectify the unreasonable risk posed by the hazardous conditions, and the company's duty extends to the protection of the workers, who were, we conclude, foreseeable victims of that danger.

We reverse the summary judgment of the district court, and hold that as a matter of law, on the record as we are required to view it at this time, Bechtel owed Caldwell a duty of due care to take reasonable steps to protect him from the foreseeable risk of harm to his health posed by the excessive concentration of silica dust in the Metro tunnels. We remand so that Caldwell will have an opportunity to prove, if he can, the other elements of his negligence action. Judgment Accordingly.

LURIA BROS. & CO. INC. v. UNITED STATES
177 Ct. Cl. 676, 369 F.2d 701 (1966)

Whitaker, Senior Judge. [The trial commissioner denied the Plaintiff's second claim for damages which was based on the allegation that it was caused by defendant's delay resulting in loss of productivity of plaintiff's labor force.]

First, it says the delay required it to work during severe winter weather between December 1, 1953, and March 10, 1954, and again between November 25, 1954, and January 19, 1955; second, from March 11 to August 11, 1954, it had to work under adverse water conditions; third, the constant revisions in the contract drawings resulted in confusion and interruption of the orderly progress of the work. For this loss of its labor, it claims damages in the aggregate amount of $131,116.66 . . . .

[The evidence on loss of labor productivity is far from satisfactory.] Plaintiff's sole witness on this point was John Crawford, who had been in plaintiff's employ for about 10 years and who was its chief of construction at the time of the work on this job was being done. However, at the time of his testimony he had left plaintiff's employ and was then chief engineer in the New York area for the Frouge Corporation, a company with which plaintiff had no connection. It was a large company with $50,000,000 worth of building construction in that area consisting of housing developments, apartments, and office buildings. Mr. Crawford had graduated as a civil engineer from Columbia University in 1924, since which time he had been engaged in both heavy and building construction work.

He was a competent witness, well-qualified to express an opinion on the loss of productivity of the labor. But, strangely, plaintiff offered no corroboration of his testimony. But, stranger still, defendant did not cross examine him on this point and offered no testimony in rebuttal. His testimony stands unchallenged.

The defendant's sole reply to plaintiff's request that the trial commissioner find the facts as testified to by this witness was in its exceptions to plaintiff's proposed findings of fact. In this document, defendant made on this reply:

Plaintiff's productivity losses, so-called were based on estimates, entirely unverifiable.

Defendant considered this item unproved. It entirely neglects allowance for time consumed by strikes, plaintiff's own delays, and other non government delays.

That loss of productivity of labor resulting from improper delays caused by defendant is an item of damage for which plaintiff is entitled to recover admits of no doubt . . . ; nor does the impossibility of proving the amount with exactitude bar recovery for the item.

It is a rare case where loss of productivity can be proven by books and records; almost always it has to be proven by the opinions of expert witnesses. However, the mere expression of an estimate of productivity loss by an expert witness with nothing to support it will not establish the fundamental fact of resultant injury nor provide a sufficient basis for making a reasonably correct approximation of damages . . . .

Crawford's testimony is unrebutted . . . . While it is the sacred duty of this court to protect the [defendant] Government from unrighteous demands as well as to protect the citizen from imposition by the Government, we cannot wholly reject this witness's testimony on the question of amount of damage.

[The court noted that witness Crawford had been an employee of the plaintiff for 10 years. Although he was not in the plaintiff's employment at the time of his testimony, the court stated that his predilection for his former employer accounted for his desire to "help them out" all he could. Although not finding that Crawford was dishonest, the court found his estimates high.]

Crawford testified that between December 1, 1953, and March 10, 1954, the productivity of plaintiff's labor force was reduced in as much as the men had to work outside on trench excavations and foundation construction in winter weather. This required them to wear gloves and warmer clothing and to work on ground which was frozen and/or extremely wet because of the rising ground water during that time of year. Based on his over-all experience in construction work and his observation of this particular job, he estimated that the average loss of productivity of labor during this period was 33 1/3%. With regard to the period between March 11, 1954, and August 11, 1954, Crawford testified that the labor productivity was reduced 25% because of adverse water conditions encountered at the construction site period. During that period from August 12, 1954, to November 24, 1954, Crawford estimated that the average loss of productivity of plaintiff's labor force was 20% because of confusion and interruption of normal job progress as a result of several revisions by defendant of the contract drawings for the [lean-to] building. Crawford also estimated that the loss during the period from November 25, 1954, to January 19, 1955, was 20% due to cold weather conditions when the men had to work in an only partially completed structure.

That winter weather and adverse water conditions reduced the efficiency of a labor force in the performance of construction work only stands to reason. It has been held by this court that when loss of productivity brought about by these conditions results from defendant's breach of contract, the plaintiff is entitled to recover its additional costs occasioned thereby as damages.

Notwithstanding the fact that Crawford estimates regarding the other three periods are unrebutted, we cannot ignore the fact that the percentages testified to were merely estimates based upon his observation and experience. Furthermore, his estimates are much higher than testified to in other cases in which the conditions were not materially different than those present here. Taking these things into consideration and in view of the fact that no comparative data, no standards, and no corroboration support his testimony, we are constrained to reduce his estimates based on the record as a whole and the courts knowledge and experience in such cases.

The total additional cost to plaintiff during the 518-day overrun period because of loss of productivity of its labor force was thus $25,031.39 .... Because of a duplication of costs included in another of plaintiff's claims, $4,550.63 must be deducted from this total figure. This results in a net cost to the plaintiff of $20,480.76 because of loss of productivity of its labor force. Since the defendant was responsible for 420 days of the 518-day overrun, plaintiff is entitled to recover 81% of this amount, or $16,589.42.

We have found that the plaintiff is entitled to recover $62,948.33 for excess home office overhead. Adding these two items to the $85,544.92 found by the trial commissioner makes a total of $165,082.67. A judgment for this amount is entered in favor of plaintiff against defendant.

DISCUSSION QUESTIONS

1. Explain the concepts of "defective plans" and "defective construction" and discuss the customary responsibilities of the design profession and the contractor with respect to the construction project.

2. What are the functions of "drawings" and "specifications"? Describe the procedures for dealing with inconsistencies between the drawings and the specifications.

3. Explain the meaning of lump-sum, unit-price, and cost-plus contracts. Discuss from the viewpoint of the owner and the contract or when such contracts are a useful technique in construction contracting.

4. Discuss the pros and cons of various types of control the construction manager may have over the general contractor. Does "as directed" and "as approved by" have the same meaning? Discuss the implications of the variants.

5. What was the result in the *Caldwell v. Bechtel case*? Did the plaintiff recover damages for the silicosis alleged to have been contracted while working in the tunnel?

6. On a job only partially completed owing to the owner's default, how should damages with respect to the contractor's anticipated profits be allocated? How should the contractor's costs be allocated with respect to employee severance pay, termination of bonds and insurance, shutdown expenses, and stop-gap work to prevent deterioration?

7. During a construction project many notices are sent or exchanged among the various parties. While some notices need not be given to the owner, the contract frequently provides that notices for the owner shall be given to the architect or the design professional. In the absence of this specific language, what is the effect of notice with respect to unanticipated subsurface conditions that was given to the architect or design professional rather than directly to the owner? This problem is illustrated by *Lindbrook Construction, Inc. v. Mukilteo School District Number 6*, 458 P.2d 1 (Wash. 1969).

11

PROPERTY
AND LAND USE

Many shades of meaning attend the term "property," but the most commonly thought of definition is cast in terms of ownership, such as referring to "my house," "my job," or "my patent." In fact, in a legal sense, the term refers to a bundle of rights including the right to possess, use, and dispose of the object.

11.1 CLASSIFICATIONS OF PROPERTY

Property concepts may be divided into several categories. The most common divisions as to its character are (1) real property, (2) personal property, or (3) fixtures. Property may also be classified with respect to its existence as (1) tangible or (2) intangible, or described in terms of its ownership as (1) public or (2) private.

Real Property and Personal Property

Real property, generally considered to be land, also includes gases, liquids, and minerals in place beneath the surface as well as the airspace above the earth and those things that are firmly attached to the land, such as bridges or office buildings. Personal property is customarily considered to be objects and rights, other than real property, capable of being owned, such as tools, trucks, and corporate stock certificates.

225

Fixtures

The term "fixture" generally refers to some object, originally personal property, but which has become part of the real property either by being attached to the soil or being attached to a structure that is legally part of the soil. Examples include large mirrors, air conditioning units, and hot water heaters.

The determination of whether a particular item is a fixture is significant since certain legal rights involving taxes, creditor's rights, and warranties may turn on the determination of whether a particular item remains personal property or has in fact become part of the real property as a fixture. If property is determined to be a fixture, it (1) passes to the purchaser upon the sale of the land unless excluded by agreement and (2) is covered by mortgages and mechanics' liens. The courts use an objective test to see whether an ordinary reasonable person would be justified in assuming that the attached item was intended to become part of the real property. Evidence may be based on custom, such as that furnaces are generally firmly attached, anticipating that they will become part of the real property, as contrasted to a TV antenna, which many owners plan to remove and take upon sale of the real property. At issue in *Strain v. Green*, 172 P.2d 216 (Wash. 1946), was a sizable plate glass mirror on the dining room wall installed by means of a large piece of three-eighth inch plywood firmly nailed to the plaster wall by 26 nails, which left large plaster damage holes when pried loose. Here, the court held that the mirror, which was in the house at the time of sale and subsequently removed by the seller, was regarded as a portion of the wall and was conveyed with the house to the purchaser under a warranty deed.

Tangible and Intangible

Classifications of tangible and intangible are primarily based on whether or not the property has physical existence. Property that physically exists such as land, machinery, and equipment is tangible. Intangible property does not have actual physical existence but rather is represented by symbolic objects such as patent rights, trademarks, shares of stock, and the right to claim damages.

Public and Private

Property, such as national parks and offshore oil and gas reserves, which is owned by the government, a political subdivision or a government agency, is referred to as public. On the other hand, private property is that owned by individuals or some form of private organization. A state university, even though supported by private endowments and charging student tuition, is nevertheless public.

11.2 FORMS OF PROPERTY OWNERSHIP

The ownership of property may be divided into a variety of ownership forms ranging from sole ownership to some type of joint ownership and may involve present, future, or limited term ownership interests.

Joint Tenancy

Joint tenancy is a form of property ownership by two or more persons with equal rights of enjoyment during their lives. Both real and personal property are capable of joint ownership, and a joint tenancy is not presumed to exist unless the intention is expressed by such words as "title to be held as joint tenants, with the right of survivorship, and not as tenants in common." A joint tenancy requires four unities (1) unity of interest, (2) unity of title, (3) unity of time, and (4) unity of possession. These four unities mean that the property of the joint tenancy must be acquired simultaneously, the title must be derived from one conveying instrument, the interest must be co-extensive, and the right of possession must be co-equal. A joint tenancy is distinguished by the "right of survivorship," by which the entire property passes to the last survivor.

Tenancy in Common

Tenancy in common is a form of ownership under which two or more persons are entitled to an undivided possession of the common property. Unlike the joint tenancy, only the unity of possession is required. The interests, however, need not be equal, and, upon the death of one of the tenants in common, the decedent's interest passes to the heirs of the deceased tenant.

Community Property

The system of community property ownership exists in several states including Florida, Louisiana, and California. This system is based on the Civil Law code as carried into those states by the Spanish and the French jurisprudence. The concept of community property rests on the theory that the husband and wife make equal contributions to the acquisition of wealth after marriage and that each spouse is entitled to share equally in the whole of the acquired property interests.

While the statutes are not uniform, they usually provide that all property acquired after marriage, unless it is by gift, devise, bequest, descent, or purchased with separate noncommunity funds, becomes community property of the husband and wife. While some statutes provide for survivorship, generally those laws provide that upon the death of one spouse, one half of the property shall go to the estate of the decedent spouse.

11.3 ESTATES OF OWNERSHIP IN PROPERTY

Real property law in the United States stems from the Norman (French) conquest of England in 1066. Estates of ownership in real property, which developed in the feudal system, may exist for periods of (1) undetermined duration and classified as a "freehold" estate or (2) a period of determined duration and classified as "less than a freehold" estate.

Freehold Estates

In medieval England, land was primarily divided between tenants who had a fee tenure (freeholds) and persons who worked for a lord and may have occupied land as serfs and did not have free tenure. A freehold estate is of undetermined duration in real property and may be either an estate in fee simple or a life estate. A fee simple estate, the strongest most absolute form of ownership in land, has been described as giving the owner "complete ownership, immediate and forever, with the right of possession from boundary to boundary and from the center of the earth to the sky, together with all lawful uses," and, upon death of the owner, the property goes to the decedent's estate. *Magnolia Petroleum v. Thompson*, 106 F.2d 217 (8th Cir. 1939).

A life estate is a freehold estate because it is of undetermined duration since it is for the life of the holder of the estate or the life of another person. The right of enjoyment and use of the life estate is somewhat more limited than that of the fee simple estate since the life estate may not convey the property and must not commit waste of the property. For example, the life estate has the right to raise crops but not sell the topsoil. This recognition of the rights of the subsequent owner, (referred to as "the remainderman"), is a doctrine that has been urged in environmental litigation. The doctrine of "waste" recognizes that the rights of the next generation may be destroyed by the excessive use of the present generation, diminishing the estate to be received by the "remainderman."

Estates Less Than Freehold

An estate in real property, that is to run for a determinable or fixed time, is "less than a freehold estate." For example, a lease for 20 years, being a fixed or determined period of time, is an estate "less than a freehold estate."

11.4 TRANSFER OF PROPERTY OWNERSHIP

Real property interests may be transferred by four methods that include sale, legal action, gift, and will or inheritance. Although it is important to have qualified counsel consider all aspects of title to real property, only the first two of these means are of particular significance to the engineer.

Transfer by Deed

Both the transfer by gift and by sale require a formal instrument of transfer. Since the transfer of land is controlled by applicable state law, the formalities of the conveyance vary from state to state.

Deeds, written instruments that transfer title to real property, are generally classified as warranty deeds and quit-claim deeds. The warranty deed contains a warranty that the title being transferred to the grantee is good. While the warranty may extend to a variety of matters, there are three essential warranties: (1) grantor has good title and the right to convey, (2) there are no encumbrances other than spelled out in the deed, and (3) assurance that the grantee and his or her heirs and assigns will have quiet and peaceful possession of the property.

A conveyance by a quit-claim deed transfers only the grantor's interest, if any, in the property and does not expressly or impliedly warrant that the grantor has any title to the land. Typically, the parol evidence rule, with its usual exceptions, applies to the written deed because the parties final expression of their agreement may not be contradicted by evidence of a prior agreement or a contemporaneous oral agreement.

Transfer by Operation of Law

Real property: Adverse possession. Title to real property may be acquired under the doctrine of adverse possession, in which case the possession ripens into a legal title. Acquiring title to real property by means of adverse possession requires compliance with the prevailing law that customarily includes a character of possession that is (1) actual and exclusive, (2) open and notorious, (3) hostile to the superior title of another, and (4) continuous for the statutory period. In addition, some states require that the claimant pay property taxes as they accrue; this is frequently referred to as color of title, since it involves an open act consistent with a claim of ownership.

Because mere intent to occupy is insufficient, the adverse claimant must physically occupy the land to meet the requirement of actual and exclusive possession. The degree of control required to satisfy this criteria is determined by the character of the land and the use for which it is adapted. For example, pasturing cattle on land ordinarily used for grazing has been held to meet this test, although some states' statutes add that the land must be fenced to give the owner notice of the adverse claim. In describing the character of the evidence essential to constitute actual adverse possession, the court requires that the proof must demonstrate that degree of possession sufficient to indicate to the public that the person claiming has appropriated the land for the exclusive use of the adverse possessor.

Personal property: Abandoned and lost. Personal property is abandoned when the property is discarded by the last owner with no intention to reclaim it. *Botkin v. Kickapoo, Inc.*, 505 P.2d 749 (Kan. 1973), defined abandonment as a state of facts demonstrating an intentional relinquishment of a known right absolutely and without reference to any particular person or for any particular purpose. In such a case, ownership of the property is acquired by the person who first exercises control over and claims title to the abandoned property.

Lost property is distinguished from abandoned property in that the lost property is mislaid as the result of negligence or some other cause. In the case of lost property, title remains in the owner who mislaid the property; however, the finder has title to the property except as to the true owner.

Bailments

A bailment exists where the owner (bailor) surrenders possession of an object of property to another person (bailee) with provision that the bailee will return the property to the owner at a later time. The elements essential for a bailment are (1) retention of ownership by the bailor, (2) possession and temporary control by the bailee, and (3) ultimate reversion of possession to the bailor unless bailor authorizes transfer to a third party.

The law of bailments and the relationship between the bailor and the bailee is determined by the conditions under which the bailee receives the property. The variant in this relationship includes (1) bailment for the sole benefit of the bailor, such as where a contractor stores materials on the land of another without the land owner receiving any compensation for its storage; (2) bailment for the sole benefit of the bailee such as a contractor borrowing a piece of equipment without paying any compensation for its use; and (3) bailment for the mutual benefit of the bailor and bailee such as a contractor using a piece of equipment that is rented from the owner.

Many courts have attempted to set different degrees of care required by the bailee based on an allocation of benefits between the bailor and bailee. That is, a bailment involving a benefit for only the bailor requires only slight care by the bailee, but, a bailment solely for the benefit of the bailee entails a high degree of care by the bailee. Where the bailment is for the mutual benefit of the parties, the bailee is held to a standard of ordinary care.

As was stated in chapter 7, a number of courts have experienced considerable difficulty in trying to resolve these disputes primarily on the basis of three distinct degrees of negligence. The purpose in attempting to make this demarcation in degrees of negligence is relevant primarily in allocating risk between bailor and bailee to assess liability or nonliability for injury or destruction of the bailed property.

11.5 LAW OF WASTE

Private Waste

The law of "waste" dates back to the 13th century in Anglo-Saxon jurisprudence. In this regard, considering the question of waste, the issue may require analysis of either private or public waste. Waste is defined as conduct, either by commission or omission, by the party in possession that results in permanent injury to the land. With respect to the matter of private waste, the law seeks to balance the expectations of the party in possession (the "life tenant") and the party not in possession (the remainderman) by assessing whether the injury to the property is of sufficient gravity to constitute permanent damage to the property. The life tenant has the right to receive the rents and profits and the duty to pay taxes and other current charges and also has the obligation to keep the premises in good repair but excludes an obligation to make permanent improvements.

Because the judicial interpretation of the term "waste" varies in the context of the property and the purpose for which the property is used, a line of cases developed a rule referred to as the "open mine doctrine," which permits the life tenant to work existing opened mines, to use established wells, and to cut a moderate amount of timber but prevents the life tenant from opening new mines, drilling new wells, or stripping the land of timber. In *Beliveau v. Beliveau*, 14 N.W.2d 360 (Minn. 1914), the defendant life tenant on a 320 acre farm was given the power of sale and the right to use the proceeds for defendant's support; however, defendant permitted the premises to depreciate in value by letting the premises decline to serious disrepair. The court concluded

> There is a community of interest between a life tenant and a remainderman which gives rise to obligations and duties as between them. By implication, a life tenant is a quasi trustee of the property in the sense that he cannot injure or dispose of it to the injury of the remainderman, even though a power of disposition . . . is annexed to the life estate.

The adaptability of the law to the development of cities and changing neighborhoods is illustrated by *Mels v. Pabst Brewing Co.*, 79 N.W. 738 (Wis. 1899), where the owner of the life estate demolished a dwelling house and graded down the lot approximately 30 feet to make it available for a business site. Noting the growth of the city, that the lot had been surrounded by factories and that the value of the lot for business use became substantially higher, the court dismissed an action for waste against the life tenant and held

> In the absence of any contract . . . to use the property for a specific purpose, or to return it in the same condition in which it was received, a radical and permanent change of surrounding conditions . . . must always be an important, and sometimes a controlling consideration on the question whether a physical change in the use of the building constitutes waste.

Public Waste

Although there is early precedent in the *Bishop of Winchester's case* in the 16th century, modern statutory law and a trend of cases have shown a real concern for the public protection against waste and especially in the field of natural resources. In that case, a lessee was enjoined from cutting down all of the timber trees, despite agreement from the remainderman, with the court noting ". . . it is against the public good to destroy the trees . . . ."

Conservation is the legal basis of state regulation over natural resources especially oil and gas production. Under general principles of property law, each property owner is entitled to drill for and produce unlimited quantities of petroleum from wells drilled on the owner's own property. In the early case of *Barnard v. Monongahela Natural Gas Co.*, 65 A. 801 (Pa. 1906)., when there was little scientific knowledge concerning the location or extent of underground petroleum, the courts used the analogy of wild animals (i.e., that gave ownership to the first person who captured or reduced a wild animal to possession) and permitted petroleum producers "unrestricted" wasteful production, reasoning that ownership was acquired by the first person to exercise control over the petroleum. However, this unrestricted production prematurely depleted the associated natural gas responsible for driving the oil production to the surface, resulting in a decrease in the ultimate quantity of recoverable petroleum. Recognizing this technology of petroleum production, the states adopted regulatory patterns setting production limits for each well in order to preserve the natural gas pressure drive of the field reservoir to assure the maximum recoverable petroleum by a slower but an extended period of production.

11.6 CONTROLS ON THE USE OF REAL PROPERTY

The State Police Power

With respect to the U.S. system of property law, which has its roots in the English Common Law, an early English legal scholar, Blackstone, wrote

> . . . nothing so generally strikes the imagination and engages the affections of mankind, as the right of property; or that sole and despotic dominion which one man claims and exercises over the external things of the world in total exclusion of the right of any other individual in the universe.

Recognizing that ownership is commonly considered as the power to exercise full authority over property, many scholars heralded Blackstone as advancing the proposition that "the regard of the law for private property . . . will not authorize the least violation of it; no, not even for the general good of the whole community." However, as a result of problems created by increased industrialization, urban growth, and population density, the mood shifted to urge that public control be exercised over the unrestrained private use of land. Accordingly, the proposition was restated that property could be used and enjoyed without control or diminution "save only by the laws of the land."

Controls on "land use" are not an innovation of the 20th century. Such controls, in various forms, existed in ancient England. In fact, the earliest Code of Roman Law (the Twelve Tables) provided for setback lines and other restrictions as early as 450 B.C. Historically, controls on "land use" in America extended back to colonial times, but the regulatory scope was somewhat limited, such as to a Massachusetts law that "prevented operation of slaughter houses in certain parts of Boston." These controls were based on state police powers "to regulate on behalf of the health, safety, morals, and general welfare of the community." On this authority, in another early case, *Commonwealth v. Tewksbury*, 11 Metc. 55 (Mass. 1846), the court sustained state action, which prevented landowners from removing sand or gravel from beach property.

Constitutional Limitations on "Taking"

The final phrase of the Fifth Amendment to the U.S. Constitution reads, ". . . nor shall private property be taken for public use without just compensation." Justice Oliver Wendell Holmes in the landmark case of *Pennsylvania Coal Co. v. Mahon*, 43 S. Ct. 158 (1922), held that a state statute, prohibiting coal mining within 150 feet of any improved property, violated the constitutional provision with respect to "taking." The plaintiff coal company had retained the coal mineral rights when selling the surface rights to an individual who then constructed a house on the purchased land. In a split five to four decision, Justice Holmes reasoned that "for practical purposes, the coal mineral rights had been retained so that the coal could be mined [arguing that] what makes the right to mine coal valuable is that it can be exercised with profit . . . and to make it impractical to mine [the reserved coal] was very nearly the same for constitutional purposes as appropriating it or destroying it." In a dissenting opinion, Justice Louis D. Brandeis pointed out that all regulation constituting an exercise of the police power places some restriction on the use of property and deprives the owner of some enjoyment of the property without compensation, referring to the Court's earlier approval of a statute that had required coal companies to leave coal pillars to support the mining roof for the protection of the miners. On this basis, Brandeis found the current restriction "merely the prohibition of a noxious use," since subsidence of the surface would destroy the house. He argued that the state had not taken the property (the reserved coal) but had merely prevented the owner from using it in a manner that interferes with the paramount rights of the public. Subsequent cases attempted to formulate a workable test of an unconstitutional "taking" by determining whether the regulated persons were required to make a disproportionate sacrifice or contribution to a public use.

The Supreme Court, in 1987, again addressed the question of "taking" in *Nollan v. California Coastal Commission*, 107 S. Ct. 3141 (1987), where the Coastal Commission denied plaintiffs a permit to demolish their run-down beach house in Ventura County, north of Los Angeles, and replace it with a new home in the same location unless the plaintiffs would grant an easement to the public to cross their property to the beach. The Coastal Commission viewed this request as an exercise of its regulatory authority over property in the coastal zone predicated on a policy to maximize public access to and use of the beach. The Coastal Commission reasoned that the purpose of the condition was an effort to enhance the public's "visual access" to the beach by overcoming the "psychological barrier" to using the beach. While the *Nollan case* provides little new guidance for determining when the critical line of unconstitutional "taking" is crossed, it suggests that governmental regulation must demonstrate a clear *nexus* between the regulation and the police power objectives.

In holding the Coastal Commission's denial of the permit as a violation of the plaintiff's constitutional rights, the Court remarked that the state is free to advance its program using its power of eminent domain but "if [the state] wants an easement across the Nollans' property it must pay for it." The Court concluded that "unless the permit's condition dedicating a right for use by the public of the Nollans' beach serves a *directly related governmental purpose*, the building restriction is not a valid regulation of land-use but an out and out extortion" [emphasis added].

First English Evangelical Lutheran Church v. County of Los Angeles, a companion case to the *Nollan case*, arose after flood waters destroyed the church's summer camp. Following the flood, Los Angeles County adopted an ordinance prohibiting any construction or reconstruction in the area of the church's camp, regardless of whether or not it subsequently was safe from flooding. Without redefining the parameters of an unconstitutional "taking," the Court held the ordinance overly restrictive and concluded that damages should be allowed since the invalidation did not compensate the landowner for the deprivation of the use of the property during the period the regulation was in effect. Described as a "temporary taking," the Court approved awarding of damages for a period ordinarily beginning when the owner seeks and is denied permission to use the land and ending when the offending regulation is declared invalid.

Obviously, the impact of the *Nollan* and *Glendale Church* cases places a new onus on the city councils or local governing bodies in the zoning and regulatory area since substantial damages may be at stake from improper action.

Zoning and Land Use

Goals of zoning. Although there have been attempts to bring the federal government into the field of land use planning, the principal power over land use controls continues to reside primarily in the states.

Zoning, the traditional technique used for urban planning, is legally sanctioned under the police power representing the state's authority to regulate for the benefit of health, morals, safety, or general welfare. Implementation of land use control is generally accomplished under state-enabling legislation, allowing local authorities to make and administer specific zoning rules.

While early land use statutes of the 1920s mainly dealt with physical characteristics specifying particular uses, such as industrial, commercial, and residential as well as height, set back, and minimum size, the objectives of zoning frequently stressed the reduction of street congestion, better fire and police protection, and promotion of health and general welfare. The Supreme Court initially declared zoning ordinances a valid constitutional exercise of the police power in *Village of Euclid v. Ambler Realty*, 272 U.S. 365 (1926), where the real estate company argued that the zoning of the company's property for residential use decreased its value and amounted to an unconstitutional "taking" without compensation. In that case, the Court declared that the zoning ordinance was presumed valid unless it was clearly arbitrary and unreasonable, having no substantial relation to the public health, safety, morals, and general welfare. This standard of judicial review, which gives great deference to legislative action, is discussed at length in chapter 14.

Historically, zoning for solely aesthetic purposes has been one of the more controversial issues. There is case law supporting zoning, which regulates architectural design and whose purpose is to protect the property values as a legitimate exercise of the police power to promote the general welfare. *Berman v. Parker*, 348 U.S. 26 (1954), *State ex rel. Stoyanoff v. Berkeley*, 458 S.W.2d 305 (Mo. 1970). Contra: *Pacesetter Homes v. Olympia Fields*, 244 N.E.2d 369 (1968). However, when the regulations deal with large residential building lots, the cases are in conflict. *Simon v. Needham*, 42 N.E.2d 516 (Mass. 1943), represents one line of cases supporting the large 3, 4, and 5 acre minimum lot sizes by pointing to fire, health, and traffic control benefits. On the other hand, another line of cases views these large residential tracts as representing invalid "exclusionary" zoning restrictions. The court in *National Land & Inv. Co. v. Kohn*, 215 A.2d 597 (Pa. 1965), struck down an ordinance that required single-family dwellings to be built on lots of at least 4 acres. In support of the ordinance, it was argued that these large lots were justified because new development would increase the population, overburden the sewage system, and increase the danger of water pollution. Finding the zoning invalid, the court rejected the contention that 4 acre zoning was necessary to prevent pollution and found that the main goal of the zoning was to prevent future burdens on the communities infrastructure by blocking the entrance to newcomers.

This bore no relation to the general welfare. *Appeal of Girch*, 268 A.2d 765 (Pa. 1970), illustrates similar results by rejecting ordinances that prevented high rise apartments.

Exclusionary zoning. To avoid some of the more troublesome problems of spot and exclusionary zoning representing a lack of coherent and comprehensive planning, states generally have constitutional or statutory provisions requiring local community compliance with specified minimum substantive and procedural standards. Essentially, these standards constitute the planning process that ultimately results in a master plan for community land use. The proper sequence is the development of the master plan followed by the more detailed zoning plan, since the zoning is merely implementation of the master plan. While states differ in the factors taken into consideration for designing the master plan, Section 65302 of the California Statutes indicates that the plan should include a statement of development policies and a diagram and a text setting forth the objectives, principles, standards, and plan proposals. In *Save El Toro Ass'n. v. Days*, 141 Cal. Rptr. 282 (Cal. App. 1977), the court interpreted these requirements as mandating that plans should contain each of the following nine statutory elements: land use, circulation, housing, conservation, open space, seismic, noise, scenic highway, and safety.

The antithesis of planning is frequently referred to as "spot zoning." Zoning that constitutes "spot zoning" is by definition invalid by arbitrarily and capriciously singling out individual units of land or small parcels for unique treatment. *Borough of Cresskill v. Borough of Dumont*, 104 A.2d 441 (N.J. 1954). To the extent that changed conditions necessitate modifications in parts of the community master plan, zoning amendments provide the most acceptable means of modifying prior zoning ordinances. These amendments are tested by essentially the same criteria as the original zoning ordinance in terms of reasonableness and the relationship between the action and the proper regulatory sphere of the police power.

Since zoning by its very nature segregates areas into single-family residential, multi-family, commercial, and industrial, the term "exclusionary zoning" would seem strangely redundant, because that term is used to define invalid classifications or zoning opposed to public uses. Perhaps the most popular use of exclusionary zoning now occurs in those situations involving claims by affordable housing advocates alleging that particular zoning requirements place unwarranted burdens on low income and moderate income people. In a leading case now known as the Mt. Laurel doctrine, the court took the position with respect to restrictions on multi-family housing units, that each community has an obligation to accept its fair share of all categories of people who desire to live in that community. Thus, town houses and multi-family housing provide people with moderate incomes "an opportunity to own their own homes." In *Southern Burlington County NAACP v. Township of Mount Laurel*, 456 A.2d 390 (N.J. 1983), the court invalidated exclusionary zoning on the grounds that "zoning ordinances must take into account not only the welfare of the residents of the community, but also the welfare of those people who live in the surrounding region [who also] contribute to the housing demands of the community." If the zoning ordinances do not provide ample area for the region's low and moderate income housing, the Mt. Laurel doctrine proclaims that such ordinances will be considered an abuse of the police power and be unconstitutional. The court reasoned that "the community is in control of the land and, in exercising that control, cannot provide housing for the rich and burden the poor through land restrictions." As an aside, the court urged the community to seek government subsidies, developer incentives, and set-aside projects.

In contrast to the Mt. Laurel doctrine, the New York courts rejected the Mt. Laurel doctrine in considering a challenge to zoning regulations, which plaintiff alleged to result in displacement of low income households from New York's Chinatown. In a two-part analysis, the court rejected Mt. Laurel and then tested the ordinance by a standard that considered whether the land use plan was "properly balanced and well ordered" and gave consideration to the "regional needs and requirements." In finding that the ordinance did not violate that standard, the court pointed to an extensive study leading up to the rezoning of the Chinatown area and was satisfied that the questions were "well considered." In addition, the court looked to empirical evidence of the city's overall record in providing lower income housing. Further, the court opined that "the housing stock in any subdistrict of the city need not be balanced as long as the city as a whole provided such housing." In a dissenting opinion, Justice John Carro criticized the majority reasoning as simplistic, denying that in a city such as New York "the constitutional requirements could not be met" by reference to "satisfying regional needs." *Asian-Americans for Equality v. Koch*, 514 N.Y.S.2d 939 (App. Div. 1987).

Judicial zoning trends. The pace and trend of the Supreme Court's thinking is indicated in two recent zoning decisions. In *City of Cleburne, Texas v. Cleburne Living Center*, 105 S. Ct. 3249 (1985), the Court rejected as invalid a zoning ordinance requiring a special use permit for group homes for mentally retarded persons but not for other group living quarters such as fraternity houses and nursing homes. The Court explained that for this zoning distinction to meet the requirements of the Equal Protection Clause of the Constitution, "the ordinance that distinguishes between the mentally retarded and others must be rationally related to a legitimate government purpose." In a later opinion, the Court upheld a zoning ordinance that prohibited adult motion picture theaters within 1000 feet of any residential zone, single or multiple family dwelling, church, park, or school. In its decision, the Court answered arguments concerning free speech guarantees by finding that the regulation was content-neutral on the time, place, and manner of speech as long as the regulation was designed to serve a substantial government interest and did not unreasonably limit alternative avenues of communication. In reaching this conclusion, the Court noted that the ordinance was aimed at the effects such adult movies have upon the surrounding neighborhood rather than the content of the movies and also noted that the restriction left more than five percent of the land area in the city open to use by such theaters.

Development impact fees. Another major issue referred to by the popular name of "development impact fees" also involves the critical constitutional questions of "taking" and stirs debate in the land development community. Development impact fees represent charges levied by municipalities on new development to finance the communities' infrastructure that would otherwise have to be paid for by the entire community through higher taxes or assessments.

Local jurisdictions have customarily been responsible for the provision of the major infrastructure and public service facilities of the community ranging from parks, schools, and roads to water and sewage facilities; but, it is common knowledge that the financing of these infrastructure items presents highly politically charged socioeconomic problems. Many communities faced with growing fiscal constraints have seized on schemes to call on subdividers and developers to contribute to the financing of off-site general improvements. These schemes have not always been cheerfully accepted, and some developers have challenged their constitutionality.

A practice of requiring the developer to build the streets and then dedicate them for public use has been generally approved by the courts as having a reasonable linkage to a public purpose. *Ghen v. Piasecki*, 410 A.2d 708 (N.J. App. Div. 1980). An extension of this philosophy that required the developer to set aside land for schools and parks was also given court approval on the rationale that not only would the residents of the new subdivision benefit from the proximity of the new improvements but also the developer would be able to charge a higher sales price and gain a greater profit because these improvements would be in place. On the other hand, since the *Nollan case* seems to have established a tight new nexus requirement that will have an analogy in the constitutional standard for development impact fees, it may be possible to argue that some fees such as highway fees are legitimate (using the terminology of "customary"), although a more serious issue may be raised in cases of a more remote fee for school or utility facilities that may be considered as too tenuous (using the terminology of "novel"). As communities devise new and more innovative means of shifting this infrastructure financial burden, the legal challenges may raise the *Nollan* thesis to place the burden on the community to explain the reasons demonstrating the nexus to the public purposes underlying the "conditions of approval."

In *New Jersey Builders Assn. v. Bernards Township*, 528 A.2d 555 (1978), the New Jersey supreme court struck down a suburban municipality's elaborate plan for apportioning and assessing the cost of new road improvements to the developers who generated the need for new roads. But a more moderate standard was adopted by the Wisconsin supreme court in *Jordan v. Village of Menomonee Falls*, 137 N.W.2d 442 (Wisc. 1965), in which the nexus is described as a reasonable connection between the development impact fee and the needs created by or the benefits conferred upon the new development.

Some insight as to how some courts may perceive the constitutionality of these fees, in the light of *Nollan* is indicated by its decision holding arbitrary and unconstitutional a development impact fee ("exaction" ordinance) that required the dedication of seven and half percent of the total acreage of a subdivision or the payment of an in-lieu fee. The courts decision was motivated, in part, by the evidence demonstrated that the town did not need or intend to use the dedicated land. See, *J.E.D. Assoc. v. Town of Atkinson*, 432 A.2d 12 (N.H. 1981). Another court phrased the attribution tests for such fees as requiring that there be

> a spatial proximity between the improvement and the [developer's] property, a correlation between the cost and the probable benefit, and an improvement capable of being exactly financed or proportioned. *Holmes v. Planning Board*, 433 N.Y.S.2d 587 (1980).

HERO LANDS CO. v. TEXACO, INC.
310 So.2d 93, (La. 1975)

Summers, Justice. Plaintiffs, Hero Lands Company (Hero), allege they are the owners of a tract of land in Orleans Parish comprising

55.5 acres. The Hero tract is contiguous to a tract belonging to Alsue Corporation, which lies to the southwest of the Hero tract in Plaquemines Parish. The common boundary of the two tracts is the Orleans-Plaquemines parish line. In March 1970, Texaco acquired an 1870 foot long and 30 foot wide right-of-way easement and servitude on the Alsue Corporation tract. This servitude runs along Alsue's boundary and is bordered by the Hero tract.

According to the petition, Texaco completed and set into operation a 24 inch high pressure gas pipeline in the servitude adjacent to and within 15 feet of the Hero property. This pipeline does not serve a public purpose which would entitle Texaco to the right of expropriation.

By not keeping its high pressure gas line at least 250 feet from the boundary of the Hero property, the petition asserts, Texaco has created and maintained a continuous dangerous nuisance affecting the Hero property. This is said to result because of the inherent hazards and dangers of such an installation, which are "now well known to the public including those purchasing land for residential or business purposes."

Further, the Hero property is alleged to be subject to the same hazards and dangers described in [a *Wall Street Journal* article.] There, an article described the circumstances surrounding the explosion of similar pipelines in Gary, Indiana, and in Coshocton, Ohio . . . . Also described are the explosion from a low-pressure system in New York in 1969 and the 1965 explosion in Natchitoches, Louisiana, which killed seventeen people.

Texaco deliberately made this installation despite its knowledge that the pipeline, in close proximity to the Hero property, would constitute a hazard. At the same time, the petition sets forth, Texaco's actions were designed to shift the damage to the Hero property and thus avoid paying Alsue the additional damage it would have incurred if the line had been located deeper within Alsue's property.

Judgment is prayed for in the amount of "at least $30,000" as damage to the Hero property by maintenance of the high pressure gas line.

Texaco raised an exception of no cause of action to this petition which was sustained by the trial judge and affirmed by the Court of Appeal. 292 So.2d 345.

The issue presented by this petition is: Does the construction of a hazardous high pressure gas pipeline adjacent and within fifteen feet of the property line separating contiguous estates give rise to an action for damage caused by this proximity which impairs the market value and full use of the neighboring estate? The question is res novo in this Court. The legal principles set forth in Articles 667, 668 and 2315 of the Civil Code are relied upon to sustain the cause of action.

Art. 667: Although a proprietor may do with his estate whatever he pleases, still he cannot make any work on it, which may deprive his neighbor of the liberty of enjoying his own; or which may be the cause of any damage to him.

Art. 668. Although one be not at liberty to make any work by which his neighbor's buildings are damaged, yet everyone has the liberty of doing on his own ground whatsoever he pleases, although it should occasion some inconvenience to his neighbor.

Thus he who is not subject to any servitude originating from a particular agreement in that respect; may raise his house as high as he pleases, although by such elevation he should darken the light of his neighbor's house, because, his act occasions only an inconvenience, but not a real damage.

Art. 2315. Every act whatever of man that causes damage to another obliges him by whose fault it happened to repair it.

As expressed in the Article, the principle is a limitation the law imposes upon the rights of proprietors in the use of their property. It is a species of legal servitude in favor of neighboring property. An activity, then, which causes damage to a neighbor property obliges the actor to repair the damage, even though his actions are prudent by usual standards; it is not the fact that the activity is carried on which is significant; it is the fact that the activity causes damage to a neighbor which is relevant.

The law, therefore, fixes the responsibility of a proprietor to his neighbor, and Texaco is a proprietor and Hero is its neighbor, within the contemplation of Article 667. Thus, the only issue presented here is whether the allegations of this petition are sufficient in law to establish that [Hero has] been damaged by the installation of the pipeline by Texaco in its servitude. If the facts alleged would, as a matter of law, constitute damage to the Hero property caused by Texaco's installation and maintenance of the line, the petition states a cause of action. For, although Texaco may use its property (servitude) as it sees fit, it can not make any work on it which may deprive Hero of the liberty of enjoying its own, or which may cause damage to Hero.

The facts considered to be true for the purpose of this exception are that the construction of the high pressure line near the boundary of the Hero property created a dangerous nuisance because the works involved inherent hazards and dangers which are well-known to the public and those purchasing land for residential and business purposes.

According to the evidence, property value is reduced within the 250 foot corridor surrounding the pipeline right of way. The hazard and danger to the Hero property is illustrated by four specific incidents involving explosions of gas lines. Also, real estate appraisers are of the opinion that such a gas pipeline depreciates the surrounding property. As a result of these facts and circumstances damage has been incurred by the Heros to the extent of $30,000.

While the owners of property are not required to suffer damage as a result of works undertaken on their neighbor's property, the law has decreed that certain inconveniences must be tolerated. Society requires this of its citizens because many lawful uses of property necessarily result in inconveniences to one's neighbors. But the extent of inconvenience the property owner must tolerate without redress depends upon the circumstances. When the actions or works cease to be inconveniences and become damaging is a question of fact. The problem is one which involves the nature of the intrusion into the neighbor's property, plus the extent or degree of damage. No principle of law confines this damage to physical invasion of the neighbor's premises, an extrinsic injury, as it were. The damage may well be intrinsic in nature, a combination of facts and conditions which, taken together, do not involve a physical invasion but which, under the circumstances, are nevertheless by their nature the very refinement of injury and damage.

For the reasons assigned, the exception of no cause of action is overruled, and the case is remanded to the trial court to be proceeded with in accordance with law.

BROWN v. MORRIS
184 So.2d 148 (Ala. 1966)

Restrictive covenants had been written in deeds in 1956 by Lookout Land Company and duly recorded. In 1961, the Browns [plaintiffs] acquired title to lots in a subdivision covered by the restrictive covenants and Morris and Pierson [defendants] acquired title to lots in the same subdivision in 1964. The Lookout Land covenants restricted the lots to "residential purposes" and buildings on each lot to "one detached single family dwelling" and also provided for minimum cost and size, setback, and minimum area requirements as to lots. The land on which the defendants were constructing their building was zoned in 1948 for business purposes and it is contended that the property upon which defendants were constructing their building, contrary to the restrictions in their deed, was zoned by the city of Gadsden for commercial use some years before the land was subdivided and the present restrictions were placed thereon.

The restrictions placed on the property by the owners when it was subdivided are more restrictive than those prescribed by the zoning ordinance. The ordinance permits the property to be used for single family residence in addition to business use. Private restrictions subsequent to valid zoning restrictions, may be more but not less restrictive. Under such circumstances, the restrictions here at issue will prevail.

It has been held by this court:

> It is well settled by the repeated decisions of this court that the owner of land, in making a sale thereof, may retain an easement or impose a servitude in the land sold, and, when not in restraint of trade, may retain in himself certain uses, which would otherwise pass to the grantee. Such retention, or limitation of the use, being a condition upon which the estate is acquired, attaches as an infirmity in the estate itself, and as a privilege or easement in the estate of the grantor, in whose favor the limitation is imposed. . . . The grantee in accepting the deed containing such conditions or covenants accepts the title encumbered thereby and is bound as though he had signed the conveyance, and he cannot complain, for he purchased and paid for only a qualified use.

Also, we observed:

> The respective rights of the parties in such premises to enforce building restrictions against another grantee is based on the fact that such scheme constitutes a part of the consideration." . . . In such cases the equitable right to enforce such mutual covenants is rested on the fact that the building scheme forms an inducement to buy, and becomes a part of the consideration. The buyer submits to a burden upon his lot because of the fact that a like burden is imposed on his neighbor's lot, operating to the benefit of both, and carries a mutual burden resting on the seller and the purchasers.

We hold that the trial court erred in denying relief to [plaintiffs] and in dismissing their bill of complaint, as amended. On remand of this case, the trial court upon finality of this opinion will forthwith enter a decree enjoining the [defendants], their agents, servants or employees, their heirs and successors, sublessees or assigns from erecting upon said property any building other than a one-family residence as provided in the restrictions and the deeds to said property, and from using said property for any other purpose.

POWELL v. TAYLOR
265 S.W.2d 906 (Ark. 1954)

George Rose Smith, Justice. This is a suit brought by six residents of Gurdon to enjoin the appellees [defendants] from establishing a funeral home in a residential district within the city. The defendants intend to remodel a dwelling known as the Taylor place and to use it as a combined residence and undertaking parlor. The plaintiffs, who own homes nearby, objected to the proposal and offered to reimburse the defendants for the preliminary expenses already incurred. This effort to the defendants having failed, the present suit was filed. The chancellor denied relief upon the ground that the neighborhood is not exclusively residential.

On this particular subject, the law has undergone a marked change in the past fifty years. Until about the end of the nineteenth century the only limitation upon one's right to use his property as he pleased was the prohibition against inflicting upon his neighbors injury affecting the physical senses. Hence the older cases went no farther than to exclude as nuisances, in residential districts, such offensive businesses as slaughterhouses, livery stables, blasting operations and the like.

Today this narrow view prevails, if at all, in a few jurisdictions only. It is now generally recognized that the inhabitants of a residential neighborhood may, by taking prompt action before a funeral home has been established therein, prevent its intrusion. In 1952 the Supreme Court of Louisiana reviewed the more recent decisions in twenty-two States and found that nineteen prohibit the entry of a mortuary into a residential area, while only three courts adhered to the older view. In a case note the matter is summed up in these words: "The modern tendency to expand equity's protection of aesthetics and mental health has led the majority of the jurisdictions to bar funeral homes or cemeteries from residential sanctuaries of ordinarily sensitive people." These decisions rest not upon a finding that an undertaking parlor is physically offensive but rather upon the premise that its continuous suggestion of death and dead bodies to destroy the comfort and repose sought in home ownership.

We have already announced our preference for the view that permits the citizens of a residential district to make timely objection to its invasion by a funeral home. In the *Fentress case*, we set aside the chancellor's injunction only because the neighborhood was changing to a business district, having already acquired drugstores, filling stations, grocery stores, etc. In that opinion we said, with reference to the proposed mortuary: "If the district of the location was an exclusively residential one, its intrusion therein would ordinarily constitute a nuisance, and could be prevented by injunction."

It is our conclusion in the case at bar that the neighborhood in question is so essentially residential in character as to entitle the appellants to the relief asked. The Taylor place is situated at the corner of Eighth and East Main Streets, and the testimony is largely directed to the area extending for two blocks in each direction, or a total of sixteen city blocks. In a relatively small city an area of this size may well be treated as a district in itself, else there might be no residential districts in the whole community. Gurdon is a city of the second class, having had a population of 2,390 in the year 1950. It is not shown to have adopted a zoning ordinance.

This square of sixteen blocks is bounded on the west by a public highway which is bordered by commercial establishments, their exact nature not being shown in detail. Otherwise the neighborhood is exclusively residential in appearance and almost so in its actual use. A seamstress living two doors east of the Taylor place earns some income by sewing at home. The couple in the house just south of the Taylor place rent rooms to elderly people and take care of them when they are ill. J. T. McAllister lives diagonally across the intersection from the Taylor place. He is in the wholesale lumber business and uses one room as an office, keeping books there and transacting business by telephone and with persons who call. A photograph of this home shows that there is no sign or anything else to indicate that business is carried on there. Farther up the street an eighty-year-old dentist has a small office in his yard and occasionally treats patients. The testimony discloses no other commercial activity within the area.

On the other hand, the residential quality of the neighborhood is convincingly shown. A real estate dealer describes it as the best residential section in Gurdon. Estimates as to the value of various homes range from $15,000 to $35,000. Many inhabitants of the area confirm its residential character and earnestly protest the entry of the mortuary. One, whose wife suffered a mental illness some years ago, says that he will be forced to move away if the funeral home is established. Another testifies that he will not build a home on his vacant lots across the street from the Taylor place if it is converted to a funeral parlor. A third testifies that she lost interest in buying the house next to the Taylor place when she learned of the defendants' plans. It is true that other witnesses state that they have no objection to the proposal, and the chancellor found that property values will not be adversely affected. But we regard the residential character of the vicinity as the controlling issue, and the evidence upon that question preponderates in favor of the appellants. [Reversed for Appellant/Plaintiff.]

Millwee, Justice (dissenting). As I read the opinion of the majority, it is now the law in Arkansas that the operation of a modest undertaking parlor in a mixed residential and business area of a city of the second class constitutes a nuisance per se and may be abated as such by injunction. This holding is so foreign to the traditional attitude of this court and the general legislative policy of this state that I must respectfully dissent.

While the majority conclude that the area in question here is "essentially" residential they proceed to apply the so-called "modern rule," which is, in those jurisdictions which recognize it, only applicable when the affected area is "exclusively" or "purely" residential. Since this goes far beyond any of the authorities cited by the majority, I suppose it should be dubbed the "ultra modern rule." It is perfectly apparent from the detailed description of the majority that the area affected here is a mixed commercial and residential one and that the chancellor was eminently correct in holding that it was not "exclusively" residential. This determination by the chancellor is in my opinion fully supported by the great preponderance of the evidence.

Moreover, the majority failed to mention the fact that appellees' contemplated operation does not include the holding of funerals or the maintenance of noisy ambulances, factors which usually accompany the operation of a funeral home. Nor does the instant case contain factors presented by the proof in the cases relied on by the majority, such as the escape of noxious odors, the depreciation of values in surrounding properties, the ability of neighbors to see the taking in and carrying out of bodies or that they will be rendered more susceptible to contagious diseases. On the contrary, it is undisputed that the structure planned by the appellees will greatly improve the beauty of the neighborhood, that there will be no noise from ambulances and no escape of odors or gases. It was further shown that within a radius of two blocks from the appellees' property there is a nursing home, a dentist office, a real estate office, a lumber office, a seamstress place of business, service station, boat factory, lumber company office, church and a hospital. Two blocks away is the business district on highway 53, and three and a half blocks away is a bulk gas plant. This could hardly be called a purely residential section.

In denying an injunction, the able chancellor stated in the decree: "The rule is well settled that no injunction should be issued in advance of the construction of a legal structure, or in advance of the operation of a legal business, unless it be certain that the same will constitute a nuisance; and, where the claim to relief is based on the use which is to be made of lawful business, the Court will ordinarily not interfere by injunction in advance of actual operation." Since the funeral home in the instant case is not a nuisance per se and may be operated in such a manner as to not be a nuisance, the rule that Chancery Courts will not issue an injunction in advance of actual operation, but will leave the complainants to assert their rights thereafter, if the contemplated use results in a nuisance, is applicable and controlling.

About the only businesses or operations which this court has seen fit to enjoin as nuisances per se are: a gaming house; bawdy house; and the standing of a stallion or jackass within the limits of a municipality. To this select group must now be added the operation of a modest undertaking parlor, where no funerals are to be held, in an area of a city of the second class which is "essentially" but not "actually" or "exclusively" residential.

It should be a matter of common knowledge that there are scores of undertaking establishments located in residential, or mixed residential and commercial, areas of the smaller municipalities in this state with hundreds of thousands of dollars invested in them. Under the rule proclaimed today these enterprises are placed in a most precarious position. And in the future, citizens such as the appellees, will be denied the privilege of pursuing a dignified and lawful calling in places where their services would be highly welcome and most sorely needed. I cannot agree to this rule.

TOPANGA ASSOCIATION FOR A SCENIC COMMUNITY
v. COUNTY OF LOS ANGELES
113 Cal. Rptr. 836, 522 P.2d 12 (Cal. 1975)

Tobriner, Judge. The parties dispute the future of approximately 28 acres in Topanga Canyon located in Los Angeles County. A county ordinance zones the property for light agriculture and single family residences; it also prescribes a one-acre minimum lot size. Upon recommendation of its zoning board and despite the opposition of petitioner, an incorporated non profit organization composed of taxpayers and owners of real property in the canyon, the Los Angeles county regional planning commission granted to the Topanga Investment Company a variance to establish a 93-space mobile home park on this acreage.

I. Scope of judicial review of administrative findings. A comprehensive zoning plan could affect owners of some parcels unfairly if means were provided to permit flexibility. Accordingly, in an effort to achieve substantial parity and perhaps also in order to insulate zoning schemes from constitutional attack, our Legislature laid a foundation for the granting of variances. The Government Code establishes criteria for these grants; it provides:

> variances from the terms of the zoning ordinance shall be granted only when, because of special circumstances applicable to the property, including size, shape, topography, location or surroundings, the strict application of the zoning ordinance deprives such property of privileges enjoyed by other property in the vicinity and under identical zoning classification. Any variance granted shall be subject to such conditions as will assure that the adjustment thereby authorized shall not constitute a grant of special privileges inconsistent with the limitations upon other properties in the vicinity and zone in which such property is situated.

Cases have held that substantial evidence must support the award of a variance in order to [e]nsure that such legislative requirements have been satisfied.

The Code clearly contemplates that at minimum, the reviewing court must determine both whether substantial evidence supports the administrative agency's findings and whether the findings support the agency's decision.

A findings requirement serves to conduce the administrative body to draw legally relevant subconclusions supportive of its ultimate decision; the intended effect is to facilitate orderly analysis and minimize the likelihood that the agency will randomly leap from evidence to conclusions. Vigorous and meaningful judicial review facilitates, among other factors the intended division of the decision making labor. Whereas the adoption of zoning regulations is a legislative function, the granting of variances is a quasi-judicial administrative one.

Moreover, courts must meaningfully review grants of variances in order to protect the interests of those who hold rights in property nearby the parcel for which a variance is sought. A zoning scheme, after all, is similar in some respects to a contract; each party foregoes rights to use its land as it wishes in return for the assurance that the use of neighboring property will be similarly restricted, the rationale being that such mutual restriction can enhance total community welfare.

Significantly, many zoning boards employ adjudicatory procedures that may be characterized as casual. Further, although we emphasis that we have no reason to believe that such a circumstance exists in the case at bar, the membership of some zoning boards may be inadequately insulated from the interests whose advocates most frequently seek variance.

II. Adequacy of planning commission findings? The variance can be sustained only if all applicable legislative requirements can be satisfied.

The acreage upon which the original real party in interest sought to establish a mobile home park consists of 28 acres; it is a hilly and in places steep parcel of land. Except for a continuous area immediately to the southeast which included an old and flood-damaged subdivision and a few commercial structures, the surrounding properties were devoted exclusively to scattered single-family residences.

The proposed mobile home park would leave 30% of the acreage in its natural state. An additional 25% would be landscaped and terraced to blend in with the natural surroundings. Save in places where a wall would be incompatible with the terrain, the plan contemplated enclosure of the park with a wall; it further called for rechanneling a portion of Topanga Creek and anticipated that the developer would be required to dedicate an 80 foot-wide strip of the property for a proposed realignment of Topanga Creek Boulevard.

The development apparently would partially satisfy a growing demand for new, low-cost housing in the area. Additionally, the project might serve to attract further investment to the region and could provide a much needed firebreak. Single-family structures apparently would necessitate costly grading, and the proposed highway realignment would require a fill 78 feet high, thereby rendering the property unattractive for conventional residential development. Moreover, the acreage is said not to be considered attractive to parties interested in single-family residences due, in the words of the report's summary of the testimony, to "the nature of the inhabitants" in the vicinity and also because of local flood problems.

The claim that the development would probably serve various community needs may be highly desirable, but it too does not bear on the issue at hand. Were that the case, a frontal attack on the present ordinance or a legislative proceeding to determine whether the area should be rezoned might be proper, but a variance would not.

Moreover, the grant of a variance for a 28-acre parcel is suspect in the absence of unusual circumstances, so large a parcel may not be sufficiently unrepresentative of the realty in a zone to merit special treatment. By granting variances for tracts of this size, a variance board begins radically to alter the nature of the entire zone. Such change is a proper subject for legislation not piece meal administrative adjudication. We conclude that the variance granted amounts to the kind of "special privilege" explicitly prohibited by the state.

DISCUSSION QUESTIONS

1. Define and illustrate real property, personal property and fixtures. Describe some legal consequences of a determination that an object belongs in one class or another.

2. What is the difference between the extent of permissible use of real property between an owner in fee simple as compared with a life tenant?

3. What is the purpose of a zoning ordinance? Under what conditions may a landowner legally object to a use restriction imposed by a zoning ordinance?

4. Is a single-family minimum area zoning regulation valid where it appears that aesthetics, in the sense of an attractive estate style neighborhood for pleasant living and to preserve the character of the neighborhood, was the underlying purpose. See, *Flora Realty & Inv. Co. v. City of Ladue*, 246 S.W. 771 (Mo. 1952) appeal dismissed 73 S. Ct. 41 (1952).

5. Does a regulation that makes future condemnation less costly constitute an unconstitutional taking where a city denied a property owner permission to plat property preparatory to development because future highway was scheduled to be built on a portion of that land? See, *Ventures in Property I v. City of Wichita*, 594 P.2d 671 (Kan. 1979).

6. Discuss the values maximized by the different opinions expressed in the zoning cases represented by Mt. Laurel doctrine and the Asian-Americans for Equality cases? Can the constitution be validly interpreted in more than one way and, if so, why?

7. Should a cement plant that constitutes a nuisance in the immediate neighborhood as the result of the emission of dirt, smoke, and vibration be permitted to continue operations by payment of damages to the adjoining neighbors? See, *Boomer v. Atlantic Cement Co.*, 257 N.E.2d 870, (N.Y. 1970).

8. Consider the validity of an ordinance that prohibits owners of private rental properties, known as "single room occupancy" dwellings (SROs), from converting or demolishing these buildings. In adopting the ordinance, the city council contended that the conversion of the SROs would contribute to the plight of the homeless by eliminating one-room residences and therefore permitted the owners of SROs to "buy out" from the ordinance for a fee of $45,000 per room.

12

INTELLECTUAL
PROPERTY

12.1 PATENTS

Patent Defined

A patent, a form of monopoly, is essentially a right granted by the government and recognized as a contract by the courts allowing the inventor (patent holder), for a limited time, to exclude everyone else from making, selling, and using the process or patented article. Because monopolies were viewed with disfavor in England, the Common Law was reluctant to grant the inventor a monopoly right to withhold from the public an invention in the form of useful and novel products or services. Subsequently, in England the sovereign began the practice of granting monopolies to inventors, and this practice of granting Letters Patent was codified in the Statute of Monopolies in 1624. The colonists brought to the United States the philosophy of patentable inventions, which was incorporated in Article I, Section 8 of the U.S. Constitution providing,

> Congress shall have the power . . . to promote the progress of science and useful arts, by securing for limited times to authors and inventors the exclusive right to their respective writings and discoveries.

Historically, Congress adopted the first Patent Act in 1790. The Patent Act, subsequently amended a number of times, is now reflected by the currently existing Patent Act of 1952 and the Patent and Trademark Office, an agency of the Department of Commerce. The right to obtain a patent is summarized in the U.S. Patent Act (35 USCA 31):

> Any person who has invented or discovered any new and useful art, machine, manufacture, or composition of matter, or any new and useful improvements thereof not known or used by others in this country, before the invention or discovery thereof, and not patented or described in any printed publication in this or any foreign country before his invention or discovery thereof, or more than one year prior to his application, and not in public use or sale in this country for more than one year prior to his application . . . may obtain a patent.

A patent assures the holder of the power to exclude anyone from the field covered by the patent, even if the same invention was developed independently and without the knowledge that the device was patented. This protection commences with the issuance of a patent presumptively establishing that the patent is valid. Since the patent is only a presumption in most patent infringement cases, the alleged infringer typically attacks the validity of the patent.

Procedure for Obtaining Patent

Under congressional authority, the Patent Office established rules of practice for patent applications. Any person, including foreign nationals (excluding a patent office employee), may be granted a patent when the application meets the statutory requirements. In the case of joint inventors, the patent may be issued to more than one inventor. However, a corporation, although it is a legal entity and considered a person for other purposes, is not deemed to be a "natural person" and, as a result, cannot be granted a patent.

A patent is a grant of a specific right to an inventor based on the agreement of the inventor to disclose the invention to the public. This right, good for 17 years, authorizes the holder to "exclude others from making, using, or selling" the invention covered by the patent. For the patent to be granted, the invention must be novel, which is interpreted to be something that is not only new or previously unrecognized but, in fact, is extraordinary. No patent will be granted if the invention is deemed to be obvious to a person of ordinary skill in the relevant art.

In *Dann v. Johnston*, 96 S. Ct. 1393 (1976), an inventor claimed a patent based on a "machine system for automatic record keeping of bank checks and deposits." This system permitted a bank to furnish customers with subtotals of various categories of transactions in connection with a single bank statement, based on a computer program after a bank customer had labeled each check by a coded category, such as "#123-food expenditures." Although the Patent and Trademark Office Board of Appeals rejected the patent claims as being anticipated by prior art, the U.S. Court of Customs and Patent Appeals (USCCPA) reversed the Board, finding that the inventor's system was narrowly drawn and "clearly within the technological arts." The Supreme Court reversed the USCCPA decision, and, in its opinion denying the patent, found that the so-called claim failed to meet the statutory test of an "invention" and failed to pass the requirement of "nonobviousness." The Court, in making this determination, stated "the criterion is measured not in terms of what would be obvious to a layman, but rather what would be obvious to a person reasonably skilled in [the applicable] art."

Since the primary purpose of the patent system is to benefit the public, the invention must be useful. The inventor must demonstrate that the invention works in practice and does, in fact, accomplish what it is supposed to do. This demonstration is referred to as "reduction to practice." Preliminary to any patent application, it is a common practice to search the files maintained by the Patent Office in Washington, D.C., to determine whether the idea of the device involved has been previously patented.

A patent will not be granted if the invention has been used in public or sold in the United States, unless the patent application was made within 1 year of the first use or sale. An additional limitation is that a patent will not be granted if the invention or discovery has been described in any printed publication in this or any foreign country either before the date of the applicant's invention or more than 1 year prior to the patent application.

A patent will only be granted when a properly completed application is properly filed with the Patent Office, together with the appropriate fee (approximately $100 to $200) based on the claim. The application must describe the device and include the inventor's name, residence, and post office address. The application (referred to as a "petition") must affirm that the inventor is "truly the originator of the device for which the patent is sought." The petition must include specifications and drawings, where applicable, and patent claims with respect to exactly what the invention is intended to accomplish and what the patent is to cover. If the patent application is rejected, the law affords the investor both an administrative and judicial appeal.

Patent Licensing: Assignment and Shop Rights

A patent may be sold or assigned in whole or in part by its owner. While a corporation may not initially acquire a patent as the inventor, it may acquire the patent by purchase and assignment. If the assignment transferring ownership is not recorded in the Patent and Trademark Office, that transfer is not valid as to subsequent assignees who later acquire an assignment without notice of the previous unrecorded assignment. The assigned rights under the patent may be fragmented with respect to the making, using, or selling the patented invention and may be restricted both geographically as well as to duration.

While the patent rights to an invention belong to the inventor, if the employee/inventor uses the employer's time and equipment in the creative activities, the law gives to the employer a limited benefit in the patents. This is referred to as a "shop right." A shop right is a nonexclusive, royalty-free, nonassignable license to use the employee's invention. In order that the employer may obtain a greater interest in the patent (providing exclusivity and assignability), many employers require that employees sign an agreement to assign patent rights to the employer.

Infringement of Patents

Any person, not owning or having a license under the patent, who "makes, uses, or sells" a validly patented invention constitutes an infringement.

Patent infringement cases, by their very nature, are lengthy, complex, and expensive. If a case is successful, the plaintiff can recover damages and, at times, compulsory royalty from the patent infringer. In some flagrant cases, the plaintiff can recover treble damages and attorney's fees.

A manufacturer of patented products must place the word "patent" and the patent number on that product. Failure to do so will bar the recovery of damages for infringement unless the infringer knew of the patent.

Infringement actions are brought in the federal district courts. In such actions, the owner may seek both damages and an injunction to enjoin further infringement, but the damages are limited to 6 years prior to the filing of the infringement action. Generally, the defense to such infringement action will be either a factual distinction between the alleged infringer's action and the patented invention or a challenge to the validity of the patent.

12.2 TRADEMARKS

Trademarks Defined

A trademark is any word, name, symbol, or the like adopted or used by and for a manufacturer's or distributor's goods to distinguish those goods from comparable goods manufactured or distributed by others. Closely related to trademarks are "service marks" to identify services. Trademarks represent "goodwill" and fall within the category of intangible property that may be sold or licensed. Trademarks must be distinctive and cannot be geographic, confusing, or descriptive of products or services.

Procedure for Obtaining Trademarks

The Common Law recognizes and protects the property right of trademarks; however, it does not provide national uniform protection such as that provided by the federal Lanham Act. This federal law, covering goods in interstate commerce, permits the owner/user to register the mark and obtain national protection. This protection, based on registration permitted after 5 years of use, requires the owner/user to file a declaration that the trademark has not been abandoned and is still in use. After such registration, the duration of this protection is 20 years and may be renewed as many times as the owner/user can demonstrate continuation of active use of the trademark in interstate commerce.

The ministerial acts of the county clerk and a secretary of state filing or reserving under the fictitious name statutes an "assumed name" or "d/b/a" for use by a business enterprise is irrelevant to trademark protection.

A trademark may be lost if the "mark," after use, becomes established as a generic term. The word "aspirin," which was once a trade name, lost that status through generic public use. Although there is a practice of indicating a registered trade name by the use of the letter "R" within a circle, that designation is not required under the law. Recent publications indicate that it currently takes an average of 13 months to obtain a federal registration, and there are more than 600,000 active registrations.

12.3 COPYRIGHTS

Copyrights Defined

Many forms of expression including books, periodicals, newspapers, photographs, reproductions of art, and architectural drawings (such as sketches, renderings, diagrams, drawings, specifications, or models), if original and creative, may be copyrighted. Under the Copyright Act of 1976 (effective in 1978), the owner of the form of expression may obtain protection from use or reproduction subject to the exemption of "fair use." This doctrine of "fair use," based on a policy to foster criticism, comments, news reporting, teaching (including multiple copies for classroom use), scholarship, or research, entitles persons to use portions from copyrighted material if the use falls within the zone of fair use. In this regard, the law explicitly permits the reproduction of multiple copies for classroom teaching purposes. While a direct copy of protected material is obviously an infringement, use of material in a slightly changed or paraphrased form may also constitute an infringement, if, under a test comparable to that used under the patent law, the material is substantially similar. The *Salinger v. Random House, Inc.*, 811 F.2d. 90 (2nd Cir. 1987) case presents an interesting fact situation in which the court blocked the publication of a biography of novelist J. D. Salinger to be based on Salinger's unpublished letters that have been donated to various university libraries. Upon learning that the biographer intended to use those unpublished donated letters, Salinger registered them for copyright and objected to publication of the biography unless references to the letters were deleted. Although the final manuscript of the biography was revised and used only paraphrased versions of the letters, the court, in finding that the language used was so close to the original letters and was so extensive and forming such an important ingredient of the book, concluded that the book constituted an infringement.

Copyright law does not protect ideas, but protects against unauthorized copying of the tangible medium or form in which ideas are expressed. Section 301 of that Act pre-empted state Common Law copyright laws and was intended to promote uniformity both by replacing state law with federal law and by eliminating frequent fuzzy questions as to when the work had become dedicated to the public. Copyrighted material should contain a notice consisting of the word "copyright" or the standard symbol "c" in a circle, the date of first publication, and the copyright holder's name.

Although the Copyright Act of 1909 provided the copyright holder protection for 28 years and the right of renewal for another 28 years, the amendments of 1976 brought the period of protection, for copyrights issued in 1978 and after, to run for 50 years after the author's death.

Comparison of Contractual and Copyright Protection

Architects or engineers have a Common Law copyright in plans and designs they have drawn and, under that copyright, are entitled to the exclusive possession and use of the work until "published" (i.e., general disclosure without claiming copyright ownership). On the other hand, prior to the 1976 Act, architectural and other design work performed as an employee or as an independent professional for hire was presumed to be owned by the employer or the person who commissioned the work. To overcome this presumption, agreements between professionals and their clients (such as architects and developers) commenced to include a stipulation that the drawings and specifications are instruments of service and remain the property of the professional. The standard American Institute of Architects (AIA) form provides "[d]rawings and specifications, as instruments of service, are the property of the architect, whether the work for which they are made be erected or not." In the absence of such a stipulation, the client ordinarily would consider that ownership of the documents was transferred to the client upon full payment of the fee to the professional. The stipulation for ownership by the design professional rests on the rationale that it would be unfair for the developer to use these plans and specifications for other projects without paying an additional fee to the professional. In contrast, if the client (developer) becomes the owner of the documents, the developer or anyone to whom the developer transfers the ownership may reuse them without fear of successful intervention by the architect/engineer. In addition, the developer or other owner of the documents could also register them and become the copyright owner of record. These contract stipulations that the drawings and specifications are only instruments of service are also consistent with the standard forms of the National Society of Professional Engineers (NSPE) stating that "the owner shall indemnify the engineer for any liability or losses resulting from reuse of the documents by the client."

While these contractual arrangements are useful between the architect/engineer and the owner/developer, it is obvious that such contract provisions, standing alone, are not effective in preventing the use of the architect's or engineer's work by a third party. That protection is only achieved under the copyright laws if the documents bear a copyright notice and are properly deposited and registered.

If the design professional's plans are used without permission for projects other than the intended one, the professional may lose not only justifiable compensation but may be exposed to liability as a potential defendant in a lawsuit resulting from reuse of the design where it is alleged that the design was negligently prepared.

Until recently, several jurisdictions held that the architect's Common Law copyright was terminated by publication upon filing the plans with the government building department. *Wright v. Eisle*, 86 App. Div. 356, 83 N.Y.S. 887 (1903), *DeSilva Constr. Corp. v. Herrald*, 213 F. Supp. 184 (M.D. Fla. 1962). However, in *Smith v. Paul*, 174 Cal. App.

2d 744, 345 P.2d 546 (1959), a California court reached a different conclusion on the theory that "it was unrealistic to attribute to the architect an intent to abandon the property rights in the plans by merely complying with a city ordinance."

Procedure for Obtaining Copyright

Copyrights are issued after an application is filed, together with the appropriate fee (nominal $10) paid to the Copyright Office of the Library of Congress. Two copies of the work to be registered must be filed with the Copyright Office. Failure to make the required filing with the Library of Congress does not affect the validity of the copyright but may result in a fine being imposed for failure to make that filing.

12.4 TRADE SECRETS

Trade secrets were protected at Common Law based on the fiduciary relationship between the parties (i.e., principal/employer and agent/-employee) by recognizing a legally sufficient property right in the person who discovered and maintained inviolate the essential aspect of the trade secret. To this end, courts exercised their equitable power to protect those property rights against attempts by unauthorized persons to convert those secrets to their own use. A trade secret was recognized as anything that, because of its secrecy, gives the holder a commercial advantage by its practice. Commonly, trade secrets are thought to include special formulas or processes for manufacturing esoteric materials, but, in fact, they also include lists of customers or qualified vendors, plant costs, wages paid, overhead, and capital resources. The property interest in the trade secret is the right to use it to the exclusion of all others, and the corollary is that the right becomes worthless without legal protection.

It is in this setting that trade secret litigation frequently arises when an employee leaves an employer either to go into a competing business or to work for a new and, frequently, competing employer. The former employer or owner of the trade secret by judicial action may seek to prevent the former employee from disclosing or using any commercial or technical information ("trade secret") belonging to the prior employer.

The RESTATEMENT of Torts defines a trade secret as

> . . . any formula, pattern, device or compilation of information which is used in one's business, and which gives him an opportunity to obtain an advantage over competitors who do not know or use it. It may be a for-

mula for a chemical compound, a process for manufacturing, treating or preserving materials, a pattern for a machine or other device . . . . A trade secret is a process or device for continuous use in the operation of the business. Generally it relates to the production of goods, as, for example, a machine or formula for the production of an article.

The subject matter of a trade secret must be secret. Matters of public knowledge or of general knowledge in an industry cannot be appropriated by one as his secret. Matters which are markets cannot be his secret. Substantially, a trade secret is known only in the particular business in which it is used. It is not requisite that only the proprietor of the business know it. He may, without losing his protection, communicate it to employees involved in its use. He may likewise communicate it to others pledged to secrecy.

Factors determining whether particular trade secrets are covered by the RESTATEMENT are (1) the extent to which the information is known outside of the business of the owner of the trade secrets, (2) the extent of measures taken by the owner of the trade secrets to guard the secrecy of the information, (3) the value of the information to the owner of the trade secrets and owner's competitors, (4) the amount of effort or money expended by owner in developing the information, and (5) the ease or difficulty with which the information could be properly acquired or duplicated by others.

GRAHAM v. JOHN DEERE COMPANY
86 S. Ct. 684 (1966)

Justice Clark. After a lapse of 15 years, the Court again focuses its attention on the patent-ability of inventions under the standard of Art. 1, Section 8 Cl. 8, of the Constitution and under the conditions prescribed by the laws of the United States. Since our last expression on patent validity, the Congress has for the first time expressly added a third statutory dimension to the two requirements of the novelty and utility that had been the sole statutory test since the Patent Act of 1793. This is the test of obviousness (i.e., whether "the subject matter sought to be patented and the prior art are such that the subject matter as a whole would have been obvious, at the time the invention was made, to a person having ordinary skill in the art to which said subject matter pertains.")

We have concluded that the 1952 Act was intended to codify judicial precedents embracing the principle [of obviousness] long ago announced by this Court in *Hotchkiss v. Greenwood*, 13 L.Ed. 683 (1851).

[This case], an infringement suit by petitioners presents a conflict between two Circuits over the validity of a single patent on a "Clamp for Vibrating Shank Plows." The invention, a combination of old mechanical elements involves a device designed to absorb shock from plow shanks as they plow through rocky soil and thus to prevent damage to the plow.

The Congress in the exercise of the patent power may not overreach the restraints imposed by the stated Constitutional purpose. Nor may it enlarge the patent monopoly without regard to the innovation, advance or social benefit gained thereby. Moreover, Congress may not authorize the issuance of patents whose effects are to remove existent knowledge from the public domain, or to restrict free access to materials already available.

Thomas Jefferson was not only an administrator of the patent system under the 1790 Act, but was also the author of the 1973 Patent Act. In addition, Jefferson was himself an inventor of great note. Jefferson, like other Americans, had an instinctive aversion to monopolies. It was a monopoly on tea that sparked the Revolution and Jefferson did not favor an equivalent form of monopoly under the new government.

As a member of the patent board for several years, Jefferson saw clearly the difficulty in "drawing a line between the things which are worth to the public the embarrassment of an exclusive patent, and those which are not."

This Court formulated a general condition of patentability in 1851 in [the *Hotchkiss* case]. The patent involved a mere substitution of materials—porcelain or clay for wood or metal in door knobs—and the Court condemned it, holding:

> [U]nless more ingenuity and skill were required than were possessed by an ordinary mechanic acquainted with the business, there was an absence of that degree of skill and ingenuity which constitute essential elements of every invention. In other words, the improvement is the work of the skillful mechanic, not that of the inventor.

The language [in that case], and in those which followed, gave birth to "invention" as a word of legal art signifying patentable inventions. Yet, as this Court has observed, "[t]he truth is, the word invention cannot be defined in such manner as to afford any substantial aid in determining whether a particular device involves an exercise of the inventive faculty."

The pivotal section around which the present controversy centers is Section 103. It provides:

A patent may not be obtained though the invention is not identically disclosed or described as set forth in Section 102 of this Title, if the differences between the subject matter sought to be patented and the prior art are such that the subject matter as a whole would have been obvious at the time the invention was made to a person having ordinary skill in the art to which said subject matter pertains. Patentability shall not be negatived by the manner in which the invention was made.

Patentability is to depend, in addition to novelty and utility, upon the "non-obvious" nature of the "subject matter sought to be patented" to a person having ordinary skill in the pertinent art.

It also seems apparent that Congress intended by the last sentence of Section 103 to abolish the test it believed this Court announced in the controversial phrase "flash of creative genius" used in *Cuno Engineering Corp. v. Automatic Devices Corp.*, 62 S. Ct. 37 (1941).

Approached in this light, the Section 103 additional condition, when followed realistically, will permit a more practical test of patentability. The emphasis on non-obviousness is one of inquiry not quality, and, as such, comports with the Constitutional strictures.

While the ultimate question of patent validity is one of law, the Section 103 condition, which is but one of three conditions, each of which must be satisfied, lends itself to several basic factual inquiries. Under Section 103, the scope and content of the prior art are to be determined: difference between the prior art and the claims at issue are to be ascertained; and the level of ordinary skill in the pertinent art resolved. Against this background, the obviousness or nonobviousness of the subject matter is determined. Such secondary considerations as commercial success, long felt but unsolved needs, failure of others, etc., might be utilized to give light to the circumstances surrounding the origin of the subject matter sought to be patented. As indicia of obviousness or nonobviousness, these inquiries may have relevancy.

The invention as limited by the Patent Office and accepted by [the inventor] rests upon exceeding small and quite non-technical mechanical differences in a device which was old in the art . . . . To us the limited claims of [the patent in issue] are clearly evident from the prior art as it stood at the time of the invention . . . . We conclude that the claims [of the patent] in issue [in the patent] must fall as not meeting the test of Section 103, since the differences between them and the pertinent prior art would have been obvious to a person reasonably skilled in that art.

APPLICATION OF THEIS
610 F.2d 786 (C.C.P.A. 1979)

Rich, Judge. This appeal is from the decision of the Patent and Trademark Office (PTO) Board of Appeals (board) affirming the rejection of [patent] claims entitled "Programmed Conversation Recording System." The claims were rejected under 35 U.S.C. Section 102(b) as drawn to subject matter which had been on sale for more than one year prior to the filing date of the patent application. We affirm.

Appellant's invention is a system capable of simulating a conversation with a respondent. The system is programmed with a series of prerecorded statements or questions which are played to the respondent. After hearing each statement, the respondent has an indefinite time in which to give a response which is recorded by the system. The system then responds to the voice or vocal pauses made by the respondent by playing the next statement or question to the respondent and recording the corresponding response until the need information has been obtained.

Appellant makes three main contentions . . . [and Appellant] disputes the finding that the systems were "on sale" prior to the critical date, although [appellant] admits that as many as six systems may have been delivered prior to that date. [Appellant] asserts that Section 102(b) cannot apply because the devices were inoperative and thus did not embody the claimed invention, and that any activity prior to the critical date was bona fide experimentation, exempting [appellant] from Section 102(b).

A. The "on sale" issue. We think the board had ample evidence by which it could properly conclude that at least one sale took place prior to the critical date. In the absence of an exception to the statutory bar, this would suffice to bar appellant from a patent. Although it is not clear from the record that delivery of the systems to Market Fax actually took place prior to the critical date, delivery is not the crucial event.

For Section 102(b) to apply, it is not necessary that a sale be consummated. It suffices that the claimed invention reduced to practice was placed on sale i.e., offered to potential customers prior to the critical date. Even if no delivery is made prior to the critical date, the existence of a sales contract prior to that date has been held to constitute an "on sale" status for the invention if it has been reduced "to a reality." [As the court said], "an invention passes out of the experimental stage and becomes a reality for purpose of the statutory bar even though it may later be refined or improved." The fact that delivery may have occurred after the critical date is of no moment. The record shows, as early as December 6, 1972, almost two months prior to the critical date, appellant was negotiating with Market Fax in an attempt to sell his systems. The Market Fax invoice does not indicate that the delivery was for an evaluation.

There was other independent evidence before the board which justifies the conclusion that the system was "on sale" prior to the critical date. Appellant's correspondence with Sears and the status reports concerning Sears and Wards, as well as the press release for the National Retail Merchants Association trade show were considered by the board to be "strong evidence" on this point. "On sale" status may be found from activity by the inventor or his company in attempting to market the invention. This is because the inventor's loss of right to a patent flows from the attempt to profit from the invention by commercial exploitation beyond the grace period allowed by Congress.

B. The experimental exception issue. Appellant argues that loss of right under Section 102(b) should not apply here because the system was the subject of bona fide experimentation prior to the critical date period. Since we have agreed with the board that appellant's system was "on sale" prior to the critical date, appellant must show that appellant comes under the *Elizabeth v. Pavement Company* experimental use or sale exception by proof which is "full unequivocal, and convincing." Appellant's arguments must fail. We agree with the analysis made by the board. The notation "6 month evaluation" on several of the invoices is equivocal. It is not clear on the face of the invoices who is to benefit from the evaluation. It is as likely that the evaluation was for the benefit of the customer, who would want to know if the system suited his purposes, as for the benefit of the appellant. Such "market testing" does not fall within the experimental use exception, because it is "a trader's and not an inventor's experiment." In addition, the notation does not appear on all of the invoices. As we have already discussed, with respect to the systems delivered to Market Fax, there were no restrictions or qualifications for the benefit of appellant. The full price was charged. Like the board, we view that transaction as an outright sale.

C. The inoperativeness issue. Appellant also relies on the theory that the bar of Section 102(b) should not apply because of, that at the time prior to the critical date when Market Fax and others were using the systems the systems were "inoperative." The theory is that during the period when the invention is not yet operative it does not exist since it has not yet been reduced to practice, and therefore cannot be used, sold, or placed on sale. We find this argument wanting because appellant has failed to show that the claimed invention was inoperative. As pointed out by the board, the chief cause of the alleged inoperativeness of appellant's systems in the hands of users was telephone line interface problems. In view of the fact that the claims do not require the use of telephones, these problems could not have prevented a reduction to practice . . . because the invention claimed is independent of telephones and can be reduced to practice without them.

DISCUSSION QUESTIONS

1. What is the theory, the policy considerations, and the legal basis on which patents are granted to inventors?

2. Discuss whether the phrase "flash of genius," which has been used to describe the work of inventors, meets the legal criteria of patentability?

3. What is the purpose of a trademark, and how is it protected? How may a trademark be lost?

4. What is the difference between trade secrets and a patent?

5. How does the doctrine of "fair use" apply to the use of copy machine reproductions of a copyrighted magazine article being used for class room study and critique by 75 students?

6. If an inventor wants to test and improve the "device" by having it tried in practice by several companies, what precautions should be taken to come within the experimental sale and use exception so as not to endanger the patentability?

7. Professor X wrote to the Y publishing company suggesting that a series of papers on earthquake research be compiled and published. After considering X's letter and conducting some market research, Y rejected the idea as not being commercial and advised X. Three years later, Y made arrangements with the authors of the research to edit and publish a compilation of the earthquake research reports. What claim, if any, does Professor X have against Y or the authors who contributed the research papers?

8. A recent newspaper article reported that a manufacturer of all-terrain vehicles alleged to contain dangerous defects, copyrighted documents required to be produced by the defendant company at the trial involving product liability. The defendant company sought a court order to prevent

the plaintiff's attorney from making any distribution of the documents. Discuss the use of the copyright laws to prevent use of those documents in a subsequent suit involving another plaintiff.

13

LABOR MANAGEMENT RELATIONS

13.1 HISTORICAL BACKGROUND

General

Labor is a major and essential factor in any engineering project, whether manufacturing or construction, that involves services and materials. One economic theory holds that all payments for goods are really only reimbursements for labor in one form or another. From the economic viewpoint of the market place, labor is merely a commodity to be priced by competitive bargaining based on supply and demand. Regardless of the theory accepted, all labor economists recognize that the performance of labor has a critical and significant impact on all construction projects. This chapter deals with the legal aspects of labor problems arising out of the labor component with respect to unionized labor in engineering projects.

This discussion of labor management relations focuses on the right of employees to organize unions and engage in collective bargaining. Absent union organization and collective bargaining, the balance of power in employment and conditions of labor are tilted toward management. Only when the two participants (labor and management) are approximately evenly balanced in bargaining power, will either party challenge the decision of the other. An individual worker finding himself or herself lacking sufficient power to question management decisions, seeks the support of fellow workers in union organization and collective strength. In its ultimate form, the worker's collective activity results in a strike (i.e., a concerted refusal to work).

Evolution of Union Organization

Under current labor law, employees have the right to form and join unions, and, through the power of union collective bargaining, employees may petition their employer for improvements in wages or working conditions.

Criminal conspiracy. Since the English Common Law proscribed anti-competitive practices as a restraint of trade, the courts held unlawful any organization of employees engaged in a similar trade for the purpose of seeking higher wages. They reasoned that rates of wages like any other commodity or other component material, must be established by free give and take bargaining in the market place without "artificially" affecting the supply of labor through the use of any non-competitive concerted employee activity. Employees who sought to organize unions for the purpose of obtaining higher wages or otherwise improving working conditions (i.e., increasing costs) were prosecuted for engaging in "criminal conspiracies." Conviction for these illegal offenses resulted in fines and imprisonment.

Anti-trust laws. In 1890, the Sherman Anti-Trust Act was passed with the objective of proscribing monopolistic business combinations and practices but leaving open the question of whether labor unions were excluded from the Sherman Act. In the Danbury Hatters case, *Loewe v. Lawler*, 28 S. Ct. 301 (1908), the Supreme Court held that a union's instigation of a boycott of retail stores that sold hats produced by a struck manufacturer violated the Sherman Act and subjected the union to treble damages. During this period, management used two methods for obstructing unionization of employees. First, the so-called "yellow-dog contract" contained a clause in the employment agreement stipulating that the employee, as a condition of continued employment, would not engage in union activities. Second, the courts granted labor injunctions to prevent anticipated irreparable damage to employers from threatened union activities.

The labor injunction granted by the courts, often *ex parte* (only after hearing the plaintiff employer's argument) became a major obstacle to union activity. The court proceeded on the premise that an employer's property included both the physical assets and also the employer's good will in relations with customers and the public and concluded that delay in filling customer's orders injured those relations. Accordingly, an employer, threatened with a strike or other economic action, could show monetary damage to good will and could ask a court for protection by an injunction. Some courts were prone to grant injunctions since they were faced with certain procedural problems and difficulties in finding union assets to satisfy damage awards.

In 1914, responding to politic pressure from labor, Section 6 of the Clayton Act exempted labor unions from the anti-trust laws. However, shortly thereafter, the Supreme Court held that this exemption only applied to those cases involving an employer and its own employees and did not extend to labor boycott situations. By 1932, during the era of the New Deal, the dramatic shift from the court anti-union sentiment to popular sympathy for unions led to passage of the Norris-LaGuardia (anti-injunction) Act. At the same time, the "yellow-dog" anti-union contract clause was outlawed, and nearly all labor injunctions were prohibited except under the most severe conditions and with strict procedural safeguards. This legislation forbade federal courts from issuing temporary or permanent injunctions in nonviolent labor disputes. Reversing the earlier Supreme Court decision, the Norris-LaGuardia anti-injunction provisions specifically applied whether or not the parties to the labor dispute stood in the proximate relationship of employee and employer. Thus, employees were free to organize, and unions could now undertake most of the familiar types of economic coercion including secondary boycotts without fear of harassing court action.

13.2 NATIONAL LABOR RELATIONS ACT

The main source of law governing the relationship between labor and management is the federal National Labor Relations Act (NLRA) or its state labor relations act clones. While there are, of course, other federal laws relating to different areas of labor, such as the Fair Labor Standards Act, Walsh-Healey Public Contracts Act, the Work Hours Act, and the Equal Pay Act, the NLRA, as amended, remains the centerpiece of labor relations.

With the adoption of the NLRA (the Wagner Act) in 1935, Congress pledged government support to unions by providing federal protection of the collective bargaining process. Section 7 of the Act, the key declaration of employee rights, reads,

> Employees shall have the right to self-organization, to form, join or assist labor organizations, to bargain collectively through representatives of their own choosing, and to engage in concerted activities for the purpose of collective bargaining or other mutual aid or protection.

The NLRA had two fundamental functions. The first was to remedy unfair labor practices that interfered with collective bargaining. Section 8 listed five forms of proscribed conduct: (1) employer interference with the rights guaranteed by Section 7; (2) employer formed or dominated "company unions"; (3) employer discrimination in hiring, firing, and other matters of employment because of union activity; (4) employer discrimination against employees because an employee testifies or files charges under the Act; and (5) employer refusal to bargain collectively with the duly established representative of the employees. The second fundamental function established administrative machinery to determine and provide employees with a means of a secret ballot election to certify (or decertify) the appropriate collective bargaining agent (union).

Only employer conduct was proscribed as unfair labor practices in the Wagner Act. The conduct of labor unions was not regulated. The Wagner Act established the National Labor Relations Board (NLRB) as the administrative agency to determine whether the federal law should be applied to a particular case, to hold elections to certify (or decertify) employee representatives, and to investigate and prevent unfair labor practices by applying sanctions for employer unfair labor practices.

With the protection afforded by the Norris-LaGuardia Act and the Wagner Act, coupled with the sympathetic attitude of government and the courts, the labor movement thrived and expanded in the next decade. Collective bargaining operating under labor-management contracts became a way of industrial life. However, with rapid growth abusive conduct goes hand in hand, and public opinion demanded legislative reform for more evenly balanced powers across the bargaining table between labor and management.

13.3 TAFT-HARTLEY ACT

By 1947, Congress was in the mood to legislate a more neutral federal labor law posture by prohibiting certain harmful union practices. To that end, Section 7 was amended to include the right of employees to refrain from engaging in concerted activities.

Under Taft-Hartley, the secondary boycott was made unlawful as were jurisdictional strikes and featherbedding. In addition, a 60 day notice of intent to strike was made a required precedent to a lawful strike. National emergency procedures were set up in an attempt to avoid national disasters and still preserve free collective bargaining. Section 301 mandated the enforcement of collective bargaining agreements in federal courts, and Section 303 provided a civil damage remedy to private parties injured by secondary boycotts.

Organized labor violently objected to the new Taft-Hartley legislation and threatened retaliatory measures against the legislators who voted in favor of its adoption. However, the unions' dire predictions concerning the union-breaking impact of the Act proved groundless, and union membership continued to grow.

The major segments of organized labor, that had earlier splintered into competing organizations represented by the American Federation of Labor (A.F. of L.), the Congress of Industrial Organization (C.I.O.), and the United Mine Workers (U.M.W.), responded to the Taft-Hartley Act by merging to strengthen politically and economically their positions.

As union membership increased, the aggregate amount of moneys collected as initiation fees, assessments, and union dues became substantial. Many union officers had only little experience and no formal training in procedures for handling those funds or meeting their fiduciary responsibilities. Congressional investigations developed evidence that some union officials maintained inadequate records and diverted funds to improper uses. These disclosures led to legal actions resulting in at least one union being put under a trusteeship, and the imposition of fines and prison sentences for officers of other unions.

13.4 LANDRUM-GRIFFIN ACT

In response to Congressional investigations, the Labor-Management Reporting and Disclosure Act (Landrum-Griffin Act) imposed new restrictions and reporting requirements on employees and labor organizations. By requiring financial disclosure, the Act sought to achieve greater responsibility and higher ethical standards of conduct by union officials. These amendments guarantee a bill of rights for employee union members, require certain financial disclosures by unions, regulate the use of "trusteeships" over local unions, and prescribe procedures for the election of union officers. By way of sanctions, the Landrum-Griffin Act differs from the Taft-Hartley Act in that it provides for criminal punishment ($10,000 fine or imprisonment up to 1 year, or both) for violations. Enforcement of the Taft-Hartley Act, on the other hand, is achieved mainly through NLRB injunctive-style cease and desist orders, which were enforced by the courts on petition of the parties.

13.5 SELECTION OF EXCLUSIVE BARGAINING
REPRESENTATIVE

In industries affecting interstate commerce, certification as the bargaining agent is largely a matter of winning a free election of eligible employees. Section 9 of the Taft-Hartley Act establishes the procedure. Certification of a union as bargaining representative begins with a petition to the NLRB for an employee election within an appropriate bargaining unit. The employees, a labor organization, or the employer may file the initiating petition.

The Meaning of Exclusivity

When the union is certified by the NLRB, the union is legally the "exclusive" bargaining representative for all the employees in the bargaining unit. An early decision, *J. I. Case v. NLRB*, 64 S. Ct. 576 (1944), involved a situation where, before there was any organization by a union, the employer offered individual employment contracts to the employees. These contracts were not mandatory, and not all employees signed them, but those that did received improved wage rates and benefits in return for the promise to serve the employer and comply with company rules. Those contracts were in effect when the NLRB certified the union as the bargaining agent. The company unsuccessfully urged that those pre-existing individual contracts barred the union's designation as the exclusive bargaining agent and that those contracts were controlling until their expiration. When the company contended that it would only negotiate presently on those matters not covered in the individual contracts, the Court found that the company unlawfully refused to bargain with the union. It analogized that since union collective bargaining agreements are similar to trade agreements, there could be no effective waiver of the benefits under such (collective bargaining agreements) trade agreements. This doctrine was further expanded by the Court in *Medo Photo Supply Corp. v. NLRB*, 64 S. Ct. 830 (1944), by holding that the employer could not bypass the bargaining agent and engage instead in negotiations about wages and other working conditions with individual employees in the plant, even though the employees initiated the negotiations and actually sought to deal directly rather than through the duly elected majority union. Such illegality of bargaining with individuals in the unit is compounded when the union representative requests to be present but is excluded. *Adolph Coors Co.*, 150 NLRB 1658 (1965).

An interesting gloss on the outer limits of protected collective activity is presented by *Emporium Capwell Co., v. Western Addition Community Organization*, 95 S. Ct. 977 (1975), in which the employer's department store was represented by a union. The collective bargaining agreement contained provisions prohibiting discrimination because of religion, sex, or age and also contained a no strike clause. A group of the employees became dissatisfied with the progress the union was making in negotiations on a grievance concerning race discrimination and started picketing in defiance of the union's refusal to approve that action. The Supreme Court found that the discharge of two protesters for holding a press conference, distributing anti-employer handbills, and engaging in picketing was not sheltered by the proviso of Section 9(a). This Section 9(a) proviso permits the employee "to present grievances and have them adjusted as long as the result is not inconsistent with the collective bargaining agreement." However, the court held that the union is charged with the responsibility of representing all employees fairly and in good faith and found that dissidents' separate protests were unnecessary.

Determining the Appropriate Bargaining Unit

Upon receipt of the petition that generally must be supported by 30 percent of the eligible employees, the NLRB must determine whether the employer's operations substantially affect interstate commerce (the basis of federal jurisdiction) and also determine the scope and composition of the appropriate bargaining unit. The statute gives the NLRB little guidance about the criteria to be employed in making unit determinations except "to assure to employees the fullest freedom in exercising their rights guaranteed by the Act." In making such determinations, the NLRB seeks to encompass a criteria to maximize an employee "community of interest." Thus, it considers questions of the desirability of grouping "skilled" and "semi-skilled" in an industrial production and maintenance unit. Frequently, more than one of several types or composition of bargaining units could be considered appropriate to represent the employees. For example, the NLRB may consider a range of options including a unit consisting of an employer-wide unit (i.e., a unit comprising all employees at several plants of a particular employer) or a unit composed of employees of a group of several employers, a unit limited to the employees of one plant, a single department, or restricted to a single or group of crafts.

The bargaining unit determination is made in such a manner as to assure employees the fullest freedom in the exercise of their rights to collective bargaining. In determining bargaining units, the NLRB gives special consideration to two groups of employees: professional employees and guards. Professional employees (engineers, accountants, etc.) are not to be included in the bargaining unit representing other employees unless the majority of the professional group votes to be included. Guards employed for protection of company property may not be included in the same bargaining unit with other employees, but the guards may have a separate bargaining unit that may not be affiliated with the union representing other employees. This decision to require separate units of guard employees is based on the reasoning that combining guard representation with other employees could partially conflict with the guards' functions. In determining the appropriateness of the bargaining unit (for example whether a craft unit should be carved out of a larger industrial unit), the NLRB will consider a variety of factors such as the history in the industry, the basic nature of the duties performed by the various employees, the degree of integration of the employers production processes, whether the craft or group of employees has maintained a separate identity, and the community of interest among the various groups of employees.

Elections to Determine Bargaining Agent

Secret ballot elections to certify the bargaining representatives are conducted by the NLRB and normally resolved by a simple majority of votes. Section 9(c)(3) provides that no election will be directed if a valid election has been held in the preceding 12 month period. In addition, the NLRB has a rule that an election normally will be precluded among employees currently covered by a valid collective bargaining agreement.

Each employee within the bargaining unit is entitled to one vote in the selection of the representative. The question as to whether employees on strike are eligible to vote was clarified in the Landrum-Griffin Act by providing that employees on a valid economic strike against their employer retain their employee status and the right to vote in an election for a year after the beginning of the strike. This right is lost, of course, if the strike is unlawful, such as a so-called illegal wildcat strike.

Finally, under the criteria of *Hollywood Ceramics Co.*, 140 NLRB 221 (1962), elections have been set aside where campaign propaganda either involved a substantial departure from the truth or was so timed as not to permit the other party to make an effective reply. In this evaluation, it is immaterial whether the misrepresentation was deliberate or not if it is determined that the campaign material may reasonably be expected to have had a significant impact on the election.

13.6 EMPLOYER UNFAIR LABOR PRACTICES

Interference with Concerted Activity

Section 8(a)(1) makes it an unfair labor practice for the employer to interfere with the employees' exercise of their rights to act in concert for mutual aid and protection. The balancing of this right with the employer's property rights and right to manage the business presents a delicate balancing of conflicting interests. In *NLRB v. Babcock & Wilcox*, 76 S. Ct. 679 (1956), the employer refused to permit nonemployee union organizers on company parking lots to distribute union literature. The NLRB, in considering the "private" property ownership aspect of the parking lot, found that this restriction violated the Act since the union organizers had no practical alternative considering the plant's physical location and emphasized the right of the employees not to "speak" but to "hear." The Supreme Court reversed the NLRB on the ground that the union had other available channels of communication and that the restriction was uniformly applied to all solicitation on company property. The decision noted that this company rule was neither discriminatorily applied nor resulted in a limitation on the employees' rights to discuss self organization.

While an employer validly may adopt rules that prohibit the employees from soliciting union memberships during working time and may restrict the distribution of literature in working areas to prevent litter, these rules may be subject to challenge as to their reasonableness. In *Republic Aviation Corp., v. NLRB*, 65 S. Ct. 982 (1945), the discharge of an employee for violation of such rules and three other employees for wearing union buttons in the plant was held to be an unlawful interference with the employees' rights. The case suggests that interference, restraint, or coercion does not turn on the employer's motive or on whether it succeeded or failed. The test is whether the employer engaged in conduct that may reasonably tend to interfere with the free exercise of employee rights under the Act. It follows that if the solicitation rule is invalid, a discharge for violation of the rule discriminates by discouraging membership in the labor organization.

The issue of an employer's denial of the union request to depart from an admittedly valid no-solicitation rule subsequent to the employer's anti-union solicitation was presented in *NLRB v. Steelworkers*, 357 U.S. 357 (1958). While the court held that the employer's denial of the union's request did not constitute an unfair labor practice, it noted,

> [T]he Taft-Hartley Act does not command that labor organizations as a matter of abstract law, under all circumstances, be protected in the use of every possible means of reaching the minds of individual workers, nor that they are entitled to use a medium of communication simply because the employer is using it.

But, the Court added by dictum,

> If, by virtue of the location of the plant and of the facilities and resources available to the union, the opportunities for effective reaching the

employees with a pro-union sage, in spite of a no-solicitation rule, are at least as great as the employer's ability to promote the legally authorized expression of his anti-union views, there is no basis for invalidating these *otherwise valid* rules [emphasis added].

Faced with an organization drive by a union, the employer under the Wagner Act was extremely restricted in what could be said to persuade the employees to join a particular union or to join none at all. The employer still is restricted in what may be said, but the limitations are much more lenient under Section 8(c) of the Taft-Hartley Act, which provides

The expression of any views, arguments, or opinion or the dissemination thereof . . . shall not constitute or be evidence of an unfair labor practice . . . if such expression contains no threat of reprisal or force or promise of benefit.

Neither the employer nor a union may use coercion or threats of reprisals in attempting to influence an employee's vote in an election to determine the collective bargaining representative. By way of illustration, the employer cannot promise that wage increases or other benefits will be given if the union loses the election. During the 24 hours immediately preceding the election, the employer must exercise particular caution. In short, the employer cannot do anything to create an atmosphere in which an election should not be conducted. The NLRB announced in *Peerless Plywood*, 107 NLRB 427 (1953) a rule that proscribes "captive audience" speeches on company time within the 24 hour period prior to the election in view of their "unwholesome and unsettling effect" and their tendency to "interfere with that sober and thoughtful choice which a free election is designed to reflect." To redress this unfair advantage of the last speaker, the NLRB will set aside the election for a violation of the rule.

The limitations imposed on the employer are not intended to inhibit the normal conduct of its business. The employer can still hire new employees, discipline, even fire an employee for cause if the disciplinary action is not motivated to discourage concerted activity including union membership. The employer can speak to the employees on company time or on their own time if attendance is voluntary. The employer can voice an opinion of unions or the result to be expected from joining a union as long as threats or promises of benefit could not be implied from those statements. The employer can state the company's legal position, the dangers, and costs of union membership and what the union can and cannot do for employees.

Employee Concerted Activity

In many cases there is no dispute concerning the "concerted activity" or whether it is protected. Typical protected concerted activity involves the discussion of unionization among employees or some concerted action with respect to wages, hours, or other conditions of employment. Employer discrimination by discharging or disciplining of an employee for engaging in such activity is an unfair labor practice. Most labor cases turn on questions of fact as to whether the employer action was based on legitimate reasons such as absenteeism, horseplay, or low productivity rather than a reprisal for an employee's union activity.

In some instances, concerted activity may occur apart from unionization. Thus, a group of employees unaffiliated with any union taking action to improve working conditions is engaged in concerted activity. In *NLRB v. Washington Aluminum Co.*, 82 S. Ct. 1099 (U.S. 1962), the Court ruled that seven employees leaving work, without permission, to protest the cold temperature in their shop was a walk-off protected by Section 7 of the NLRA.

It should be noted that the protected status of collective bargaining may be lost by employee misconduct such as violation of the no strike clause of the contract. *NLRB v. Local IBEW (Jefferson Standard Broadcasting Co.)*, 74 S. Ct. 172 (1953) presents an interesting fact situation in which the employees engaged in picketing and handbilling in order to pressure the employer to grant their economic demands. The handbills and other publicity attacked the quality of the employer's broadcasting without mentioning there was a labor dispute. The Supreme Court held that the employees were not engaged in protected activity, reasoning that the employer could discharge the employees for insubordination, disobedience, or disloyalty although an employer could not use any of these grounds where the real reason was to retaliate against the employee for engaging in protected concerted activity.

This protection extends to conduct threatening to withdraw or promising to bestow benefits on employees to interfere with an incumbent bargaining representative or an employee organizational effort. "The danger inherent in a well-timed increase in benefits is a suggestion of a fist inside the velvet glove [and] employees are not likely to miss the inference that the source of the benefits conferred is also the source from which future benefits must flow and which may dry up if it is not obliged." *NLRB v. Exchange Parts, Inc.*, 85 S. Ct. 457 (1964).

Primary Strikes as Concerted Activity

The NLRA specifically preserves a union's right to strike, but there are limitations on this right. In turn, the limitations depend on the objective sought, the means used, and whether the collective bargaining agreement contains a no-strike provision.

In the absence of a no-strike provision in the labor contract, which is a commitment not to strike during the term of the contract, a union's resort to strike may be quite lawful. In that case, the strikers' rights depend largely upon the objectives sought to be achieved by the strike. A so-called economic strike is one in which the employees seek an economic objective such as an increase in wages, improvement of working conditions, a change in overtime policy, or some other such concessions from the employer. An unfair labor practice strike is an employee response to an unfair labor practice being committed by the employer and which seeks to force the employer to redress the unlawful practice.

Economic strikers remain employees, but they may be replaced by their employer. However, a discharge of employees for engaging in strike activity constitutes an unfair labor practice by "discriminating against collective action." If the employer hires permanent replacements before the economic strikers indicate a desire to return to work, the strikers, while entitled to reinstatement, are not entitled to replace the newly hired employees.

Strikers participating in an unfair labor practice strike occupy a stronger legal status. If they are replaced by other employees, the strikers retain their rights to reinstatement even though the replacement workers may have to be discharged to make a vacancy. Of course, strikers may engage in conduct during the strike that will justify disciplinary action apart from the strike. Misconduct by the strikers, such as violence or threats of violence, could be cause for disciplinary action.

Duty to Bargain

The ultimate purpose of unionization is to obtain a collective bargaining agreement under which employment conditions, which include wages, working hours, employee benefits, grievance procedure, and arbitration, are contractually guaranteed. To achieve that goal, Section 8(d) of the Act provides

> For the purposes of this section, the duty to bargain collectively is the performance of the mutual obligation of the employer and the representatives of the employees to meet at reasonable times and confer in good faith with respect to wages, hours, and other terms and conditions of employment, or the negotiation of an agreement, or any question arising thereunder, and the execution of a written contract incorporating any agreement reached if requested by either party, but such obligation does not compel either party to agree to a proposal or require the making of a concession.

To encourage the parties to negotiate toward that end, the law provides that it is an unfair labor practice for either party (employer 8(a)(5) and union 8(b)(1)(B)) to refuse to bargain in good faith with the other. Bargaining simply means meeting at reasonable times and conferring in good faith on employment-related subjects. Dilatory tactics, by protracting the interval between or in the conduct of the bargaining meetings such as six meetings of relatively short duration (roughly 3 hours) over a 6 month period, were held in *Insulating Fabricators, Inc.*, 338 F.2d 1002 (4th Cir. 1964), to violate the duty to bargain. One court suggested that in determining the reasonableness of the bargaining effort, the party's conduct may be tested against comparable conduct in negotiations for a business contract or a bank loan that the parties are desirous of concluding.

It has been held that the employer violates Section 8(a)(5) when it insists that negotiations cease until there is a change in the composition of the negotiating team selected by the bargaining representative. A similar freedom to bargain through representatives of its own choosing is expressed in Section 8(b)(1)(B), which makes it a union unfair labor practice for the union "to restrain or coerce [the employer] . . . in the selection of [the employer's] representatives . . . ."

If the employer is in possession of information relevant to the union in connection with the bargaining process, the employer must turn over that information to the union. In *NLRB v. Truitt Mfg. Co.*, 351 U.S. 149 (1956), the company contended that it was financially unable to meet the union's wage demand of a 10 cent per hour increase and that any increase in excess of 2.5 cents would put the company out of business. The company rejected the union's request to have a Certified Public Accountant examine the company's books to verify the company's capacity to meet the wage demand. In holding that this failure to supply the requested information was a refusal to bargain, the court stated,

> Good-faith bargaining necessarily requires that claims made by either bargainer should be honest claims. If such an argument is important enough to present in the give and take of bargaining, it is important enough to require some sort of proof of accuracy.

Section 8(d) provides that the duty to bargain in good faith does not "compel either party to agree to a proposal or require the making of a concession." This statutory statement concedes that "good faith" bargaining is consistent with firmness and an unwillingness to yield a position that can be backed by economic staying power in the event of a strike or lockout. However, "bad faith" may be found when the employer proposes and insists upon a set of proposals that are predictably unacceptable.

Union Security

Unions traditionally sought to obtain a contract provision that required employee membership in the union and collected the union dues by payroll deductions (check-off). Since the passage of the Taft-Hartley amendments in 1947, it has been unlawful to form any new closed shop under which employees must be union members before they are hired which, prior to the prohibition, had the resulting effect that the union rather than management selected prospective employees. Under the closed shop, if an employee proved unsatisfactory, the company had little or no right to replace the employee; generally, the company was forced to keep whatever the union sent. Since 1947, the only lawful new form of agreement to assure union membership of employees is the so-called union shop. In a union shop, the employer hires new employees, and each employee has a period of at least 30 days (except for building construction industries) to decide whether to join the union. However, once the employee has joined the union, the employee must continue membership for the life of the collective bargaining agreement.

The Taft-Hartley Act effectively restricted union security in two ways: (1) The employer was permitted to hire persons other than union members (although, under the union shop, they could be required to join the union within 30 days), which eliminated union control over the referral of job applicants, and (2) the employee after being hired need not, as a condition of retaining employment, attend union meetings, and refrain from criticizing union representatives but need only "tender the periodic dues and initiation fees uniformly required" of union members (generally referred to as an "agency shop").

Management, Its Agents and Representatives

In the context of the employer's responsibility, it is necessary to consider who may bind the employer in the commission of unfair labor practices. The term "employer" includes the management of a company, not only the president and general manager but also salaried line or staff employees with supervisory functions who are not included in the bargaining unit. Thus, what a foreman, an engineer, or a production control clerk says or does prior to an election may be interpreted as the words or deeds of the employer.

13.8   UNION UNFAIR LABOR PRACTICES

Union Interference, Restraint, and Coercion

Section 8(b)(1) provides

> It shall be an unfair labor practice for a *labor organization or its agents*
> (1) to restrain or coerce (A) employees in the exercise of the rights
> guaranteed in section 7; provided, that this paragraph shall not impair
> the right of a labor organization to prescribe its own rules with respect to
> the acquisition or retention of membership therein; or (B) an employer in
> the selection of his representatives for the purposes of collective bargain-
> ing or the adjustment of grievances [Emphasis added].

Just as when an employer engages in conduct toward employees
violating their freedom of choice of collective bargaining representative
and other concerted activity, so union conduct, which forces or intim-
idates employees in their freedom of choice, violates Section 8(b)(1) of
the Act. A threat by the union to nonstrikers or an order not to cross a
picket line accompanied by a threatening gesture is sufficient evidence
of union interference with the employees' rights. While the NLRB has
not ordered the union to pay compensatory damages (back pay) to the
employee victims, the union may be subject to several sanctions. If the
action involves violence, (1) the union may be subject to an unfair labor
practice and an NLRB cease and desist order, (2) the union and its
agents may be subject to state court criminal prosecution, and (3) if the
individuals engaging in the violence are employees, they may be subject
to disciplinary action.

Section 8(b)(4)(B): Secondary Boycotts

Section 8(b)(4), one of the most complex concepts of labor law, is
generally meant to restrict practices referred to as "secondary boycotts."
Rather than declare illegal the "secondary boycott" in express terms,
the Act recognizes as a union unfair labor practice, conduct by which a
labor organization induces

> any employees to engage in a strike or a concerted refusal . . . to use,
> manufacture, . . . handle or work on any goods . . . or to perform any ser-
> vices where an object thereof is forcing . . . any employer or other person
> to cease using, . . . handling, transporting, or otherwise dealing in the
> products of any *other producer* . . . or to cease doing business with any
> *other person* . . . [emphasis added].

The typical fact situation, similar to the Danbury Hatters case
described at the beginning of this chapter, involves the employees of

employer A, where a current labor dispute exists, seeking to bring added economic pressure on employer A by encouraging employer B (who may be a supplier or customer of employer A) to cease doing business with employer A. The pressure on employer B generally results from picketing employer B's premises.

In congressional debates, Senator Robert A. Taft remarked that this language "makes it unlawful to resort to a secondary boycott to injure the business of a third party who is wholly unconcerned in the disagreement between the primary employer and his employees." This union unfair labor practice requires evidence of a proscribed *means* (a work stoppage or an inducement to engage in a work stoppage among secondary employees) for a proscribed *object* (to force the secondary company) to cease doing business with the primary employer. However, this literal description is too broad since it would encompass some conduct that is the normal object of a primary strike (the pickets at the primary company normally seek to encourage the employees of all other employers, including secondary companies, to refrain from crossing the picket line, which is a lawful activity). *NLRB v. International Rice Milling Co.*, 341 U.S. 665 (1951). Picketing at the premises of a secondary company for that purpose, however, would be within the sweep of the prohibition. *NLRB v. Local 825, Operating Engineers*, 400 U.S. 297 (1971).

Section 8(b)(4) also contains several exceptions and provisos. One exception states "nothing in the clause is to be deemed to make unlawful any primary strike or picketing," and one proviso reads "it shall not be a violation for any person to refuse to enter the premises of an employer other than his own if the employees of that other employer are engaged in a strike ratified or approved by a union representing those employees."

Ally doctrine. If the struck primary employer (employer A), in order to continue operations, seeks out the services of another company (employer B), the primary employer cannot claim that the other company is neutral in the labor dispute. In *Douds v. Metropolitan Federation of Architects*, 75 F. Supp. 672 (S.D.N.Y. 1948), the employees of a company (company A) supplying engineering services went on strike to support demands for a new contract. During the strike, the corporation substantially increased the amount of work that it had previously contracted out to another company (company B) and sent its supervisory personnel to that company to oversee the subcontracted work. It was held that picketing of the company (company B) that accepted the subcontract work was not within the scope of Section 8(b)(4)(B), since neutral status of the company accepting the subcontracted work was lost and as an ally was subject to the same economic pressure as the primary employer. The ally doctrine has also been applied where two companies are deemed to be alter egos because of their common ownership or common control. This "integrated enterprise" theory was applied in *NLRB v. Denver Building and Construction Trades Council* 71 S. Ct. 943 (1951), in which unions struck a general contractor on a construction job site because the general contractor had awarded a subcontract for electrical work to a company employing nonunion labor. The strike had the effect of closing down the entire project. The Supreme Court found that this constituted an unlawful secondary boycott, rejecting the union's argument that their dispute was only with the general contractor and their picketing was only to force the general contractor to have an all union job. It was clear that if all the employees had been employed by the general contractor, the picketing would not have been a violation; but, here, the picketing was intended to require the general contractor to terminate dealing with the subcontractor.

Common situs picketing. Closely related to the ally question is the issue of common situs picketing. A construction project frequently may have a number of contractors performing various parts of the work. Some contractors may have union contracts, and others may use employees of another union or nonunion labor. A union engaged in an economic struggle with one contractor (company A) can maximize its bargaining power if it can shut down the entire project by posting picket signs that other employees (employees of employers B and C) will refuse to cross. To avoid this problem, the law permits the prime contractor to set up separate gates for other contractors (employers B and C) so that the lawful picketing can be at the gate of the primary employer (company A) involved in the dispute and the employees of the remaining employees will not be splattered in the fray. These rules, which give consideration to the Act's protection from secondary boycotts, are intended to minimize the injury to the neutral contractors without depriving the primary contractor's employees of their rights.

NLRB v. LOCAL 3, INTERNATIONAL BROTHERHOOD
OF ELECTRICAL WORKERS
317 F.2d 193 (2nd Cir. 1963)

Anderson, District Judge. Picoult was awarded a contract by the General Services Administration to renovate the Federal Post Office building in Brooklyn. Local 3 [protested] the award of the contract to Picoult, which chose to deal with a different union for electrical employees. Local 3 [picketed] the building, including side and rear delivery areas which were not used by the general public. On two occasions, deliveries to Picoult by employees of other companies were turned away by the pickets. Local 3 claimed that the object of its picketing this job was to have the subcontract let to a company which recognized it as bargaining representative and if they were unsuccessful in that objective, then they sought to oust Picoult. The Board found the picketing violated Section 8(b)(7)(C).

Section 8(b)(7)(C) reads in part as follows:

> . . . to picket . . . or threaten to picket . . . any employer where *an object thereof* is forcing or requiring an employer to recognize or bargain with a labor organization . . . or forcing or requiring . . . the employees of an employer to accept or select such labor organization as their collective bargaining representative, unless such labor organization is currently certified as the representative of such employees:
>
> (A) where the employer has lawfully . . . recognized any other labor organization . . .
> (B) where within the preceding twelve months a valid election . . . has been conducted
> (C) where such picketing has been conducted without a petition being filed under section 9(c) [but provided] nothing in this subparagraph (C) shall be construed to prohibit picketing or other publicity *for the purpose of* truthfully advising the public (including customers) that an employer does not employ members of, or have a contract with, a labor organization [emphasis added].

One of the principal difficulties in construing and applying subparagraph (C) is that Section 8(b)(7) contains the partially synonymous words, "object" and "purpose," used in two distinct contexts but to which much of the same evidence is relevant. These are: "where an object thereof is forcing or requiring an employer to recognize or bargain . . ." and "for the purpose of truthfully advising the public . . . ." It does not necessarily follow that, where an object of the picketing is forcing or requiring an employer to recognize or bargain, the purpose of the picketing, in the context of the second proviso, is not truthfully to advise the public. The union may legitimately have a long range or strategic objective of getting the employer to bargain with or recognize the union and yet under the proviso the picketing may be permissive. This proviso gives the union freedom to appeal to the unorganized public for spontaneous popular pressure upon an employer; it is intended to exclude the invocation of pressure by organized labor groups or members of unions, as such.

The permissible picketing is, therefore, that which through the dissemination of certain allowed representations, is designed to influence members of the unorganized public, as individuals, because the impact upon the employer by way of such individuals is weaker, more indirect and less coercive.

In this connection what is meant by "advising the public" as used in the second proviso, is highly pertinent. Congress expressly provided that the word "public" should not be so narrowly construed as to exclude consumers, but the whole context of the phrase in which it appears makes it clear that it was not intended to be so broadly defined as to include organized labor groups which, at a word or signal from the picketeers, would impose economic sanctions upon the employer; otherwise Section 8(b)(7) would be, in effect, almost entirely emasculated. By this latest amendment to the Taft-Hartley Act, Congress sought to circumscribe a kind of picketing, which by its nature, could in most cases bring an employer to his knees by threatening the destruction of his business and which, because of the attendant loss of employment, had a material tendency to coerce employees in their freedom to accept or reject union membership or freely select the union they wanted to represent them.

Professor Cox of Harvard, now Solicitor General, who worked with the Senate Labor Committee Chairman on the Section 8(b)(7)

> Picketing before a union election is divided by section 8(b)(7) into two categories: (1) picketing which halts pick-ups or deliveries by independent trucking concerns or the rendition of services by the employees of other employers, and (2) picketing which appeals only to employees in the establishment and members of the public. . . . The theory is that the former class of picketing is essentially a signal to organized economic action backed by group discipline. Such economic pressure if continued, causes heavy loss and increases the likelihood of the employer's coercing the employees to join the union. In the second type of picketing, the elements of communication predominate. If the employer loses patronage, it is chiefly because of the impact of the picket's message upon members of the public acting as individuals." The Landrum-Griffin Amendments to the National Labor Relations Act. 44 *Minnesota Law Review* 257.

Although the two categories are described by him in terms of the *effect* of each, the express language of the second proviso uses the words "for the purpose of " and it is difficult to see how they can be ignored. Nevertheless, the description of the two categories is helpful in gaining insight to the second proviso. The concepts of "signal" picketing and "publicity" picketing should be used in characterizing the union's tactical purpose rather than in describing the picketing's effect. Yet, purpose can be determined only through what is said and done under certain circumstances . . . . [T]he effect of the picketing is one of the circumstances considered in determining in any case what the purpose was in so far as it is the natural and logical consequence of what the picketeers are saying and doing.

The effect might fall short of "inducing any individual employed by any other person in the course of his employment, not to pick up, deliver or transport any goods or not to perform any services" and still be evidence of non-permissive purpose, such as display of qualifying signs accompanied by hostile gestures; or speech directed to persons unconnected with organized labor and not employees of secondary employers, such as a casual passer-by; or, for example, by forming a shoulder to shoulder picket line across an entrance which affected only members of the unorganized public who were not employees of a secondary employer.

Under the [proviso] it is the difference in purpose which determines which is permissible picketing and which is not. In its context the [proviso] means in terms of "signal" and "publicity" picketing that while most picketing with a "signaling" purpose is proscribed, most picketing for publicity is protected; the exceptions are that signal picketing is permissible when an object thereof is not forcing or requiring an employer to recognize or bargain, and publicity picketing is proscribed when it communicates more than the limited information expressly permitted by the [proviso] or it is apparently the purpose to advise organized labor groups or their members as shown by signal effects, unless there is persuasive proof that those effects are inspired by the employer who is seeking thereby to prevent legitimate [proviso] picketing by the union.

The Board must, therefore, approach its conclusion as to whether or not the picketing was "for the purpose of truthfully advising the public" by way of a finding of whether or not the union's tactical purpose was to signal economic action, backed by organized group discipline. Accordingly the case is remanded.

FRANKS v. BOWMAN TRANSPORTATION CO.
424 U.S. 747, 96 S. Ct. 1251 (1976)

This case presents the question whether identifiable applicants who were denied employment because of race after the effective date and in violation of Title VII of the Civil Rights Act of 1964 may be awarded seniority status retroactive to the dates of their employment applications.

Petitioner Franks brought this class action in the U.S. District Court of Georgia against his former employer, respondent [Bowman Transportation Company] and [International Union of District 50] alleging variously racially discriminatory employment practices in violation of Title VII. [Title VII was enacted to "assure equality of employment opportunities by eliminating those practices and devices that discriminate on the basis of race, color, religion, sex or national origin."] Petitioner Lee intervened on behalf of himself and others similarly situated, alleging racially discriminatory hiring and discharge policies limited to Bowman's employment of "over-the-road" (OTR) truck drivers.

[The District Court found that Bowman engaged in a pattern of racial discrimination in various policies, including hiring, transfer, and discharge of employees and found that the discriminatory practices were perpetrated in Bowman's collective bargaining agreement with the unions. However, the District Court declined to grant to the unnamed members of Class 3 claims (those represented by Petitioner Lee) any specific relief sought which would have included an award of back pay and seniority status retroactive to the date of individual applications for an OTR position.]

[The Court of Appeals] in affirming the District Court's denial of seniority relief to the Class 3 group of discriminatees, held that the relief was barred by Section 703(h) of Title VII. We disagree. Section 703(h) reads:

> Notwithstanding any other provision of this Title, it shall not be an unlawful employment practice for an employer to apply different standards of compensation, or different terms, conditions, or privileges of employment pursuant to a bona fide seniority or merit system . . . provided that such differences are not the result of an intention to discriminate because of race, color, religion, sex or national origin . . . .

The Court of Appeals reasoned that a discriminatory refusal to hire "does not affect the bona fides of the seniority system." Thus, the differences in benefits and conditions of employment which a seniority system accords to older and newer employees is protected as not "an unlawful employment practice by Section 703(h)." [This is] . . . clearly erroneous.

The underlying legal wrong affecting [the Class 3 group] is not the alleged operation of a racially discriminatory seniority system but of a racially discriminatory hiring system. Petitioners do not ask modification or elimination of the existing seniority system, but only an award of the seniority status they would have individually enjoyed under the present system but for the illegal discriminatory refusal to hire. It is this context that must shape our determination as the meaning and effect of Section 703(h).

On its face, Section 703(h) appears to be only a definitional provision . . . . [That section] certainly does not purport to qualify or proscribe relief otherwise appropriate under the remedial provisions of Title VII Section 706(g) in circumstances where an illegal discriminatory act or practice is found.

[An interpretative memorandum in the *Congressional Record*] stated that "Title VII would have no effect on established seniority rights. Its effect is prospective and not retrospective."

Accordingly, whatever the exact meaning and scope of Section 703(h) in light of its unusual legislative history and the absence of the usual legislative materials, . . . it is apparent that the thrust of the section is directed toward defining what is and what is not an illegal discriminatory practice in instances in which the post-Act operation of a seniority system is challenged as perpetuating the effects of discrimination occurring prior to the effective date of the Act. There is no indication in the legislative materials that [Section 703(h)] was intended to modify or restrict relief otherwise appropriate once an illegal discriminatory practice occurring after the effective date of the Act is proved—as in the instant case, a discriminatory refusal to hire. We therefore hold that the Court of Appeals erred in concluding that, as a matter of law, Section 703(h) barred the award of seniority relief to the Class 3 members.

There remains the question whether an award of seniority relief is appropriate under the remedial provision of Title VII, specifically, Section 706(g).

The legislative history supporting the 1972 amendments of Section 706(g) . . . affirms the breadth of this discretion:

> the provisions of that section are intended to give the courts wide discretion exercising their equitable powers to fashion the most complete relief possible. The Act is intended to make the victims of unlawful employment discrimination whole and . . . the attainment of this objective . . . requires that persons aggrieved by the consequences and effects of the unlawful employment practices be, so far as possible, restored to a position where they would have been were it not for the unlawful discrimination.

Seniority standing in employment with respondent Bowman computed from the departmental date of hire, determines the order of layoff and recall of employee. Further, job assignments for OTR drivers are posted for competitive bidding and seniority is used to determine the highest bidder. As OTR drivers are paid on a per-mile basis, earnings are therefore to some extent a function of seniority. Additionally, seniority computed from the company date-of-hire determines the length of an employee's vacation and pension benefits. Obviously to merely require Bowman to hire the Class 3 victim of discrimination falls far short of a "make whole" remedy. . . . Without an award of seniority dating from the time at which he was discriminatorily refused employment, an individual who applies for and obtains employment as an OTR driver pursuant to the District Court's order will never obtain his rightful place in the hierarchy or seniority according to which these various employment benefits are distributed.

Bowman raises an alternative theory of justification. Bowman argues that an award of retroactive seniority to the class of discriminatees will conflict with the economic interests of other Bowman employees. Accordingly, it is argued, the District Court acted within its discretion in denying this form of relief as an attempt to accommodate the competing interests of the various groups of employees.

We reject [Bowman's] argument for two reasons. First, the District Court made no mention of such considerations in its order denying the seniority relief. As we noted [in *Albemarle Paper case*, 95 S. Ct. 2362, 2372, (1975)],

> . . . if the District Court declines due to the peculiar circumstances of the particular case to award relief generally appropriate under Title VII, it is necessary . . . that . . . [the court] carefully articulate its reasons" for so doing. Second, and more fundamental, it is apparent denial of seniority relief to identifiable victims of racial discrimination on the sole ground that such relief diminishes the expectations of other, arguably innocent, employees would if applied generally frustrate the central "make-whole" objective of Title VII.

[Editor's Note] Justice Powell, joined by Justice Rehnquist, agreed with the majority that Section 703(h) was not in all instances a bar to the award of retroactive seniority. However, these two justices asserted that the Court erred in not recognizing a distinction between "benefits" type seniority and "competitive" type seniority. In the latter type, Justice Powell concluded: ". . . there should be no presumption favoring the retroactive granting of seniority. The District Court should not be precluded from considering the impact of such a remedy upon innocent employees, particularly since the employer [Bowman] who has discriminated is not directly affected by such an award."

DISCUSSION QUESTIONS

1. Describe the NLRB procedure available for a union to be certified as the collective bargaining agent of a company's employees.

2. What is the meaning of the following terms: (1) strike, (2) refusal to bargain, (3) yellow-dog contract, (4) closed shop, (5) labor injunction, and (6) agency shop.

3. During a strike the union seeks to prevent employees from breaking ranks and returning to work by disciplining "scabs." Discuss such action and its potential for violating the NLRA.

4. Discuss the employer's right to discipline employees and limitations on that right under the NLRA.

5. Faced with a threat of unionization, what steps may an employer take to put forward the employer's position to its employees? What restrictions does the NLRA place on the employer in discouraging employees from joining a union by (1) restricting the union's access to employees on company property, (2) prohibiting the dissemination of union literature on company property and during working hours, and (3) granting or denying wage increases?

6. In the *Franks case*, Title VII is said to prohibit employment based on certain discriminatory classifications including sex. Is it proper to advertise or select construction workers who have to install heavy steel by excluding female applicants based on their generally lesser ability to do heavy lifting?

7. Employer X has experienced at its plant a number of sporadic work stoppages by the union. During the current negotiations of a new labor contract, X seeks to obtain a "no-strike clause" in the contract, but the union refuses to discuss the matter, insisting that the right to strike is necessary to deal with employee problems that arise during the term of the contract. Can X file a charge of any unfair labor practice with the NLRB and with what result?

14

THE ADMINISTRATIVE PROCESS

14.1 ADMINISTRATIVE GOVERNMENT

Growth of Administrative Functions

Administrative regulations and sanctions are major factors influencing engineering activities. Experienced contractors recognize that many aspects of their construction and manufacturing projects, ranging from the vast military and space complex to local libraries and waste disposal facilities, are managed or regulated by a federal or state agency or some public body at the local level. Administrative agencies are responsible for licensing contractors and registration of professionals. In addition, other agencies have duties for a wide spectrum of matters such as setting standards for and performing inspections of workplace safety, achieving environmental goals, and implementing land use controls. Administrative agencies generally were established to do work of the government (i.e., promulgate rules and issue orders) in a simpler more direct fashion than if the legislature did the job by enacting a law and the courts applied it on a case-by-case basis.

Rule making is not a new concept. The promulgation of general regulations has been a normal legislative function since the government was established. The first Congress provided that traders with Indians should be licensed and bonded to observe "such rules, regulations and restrictions . . . as the president shall prescribe." Thereafter, during the

19th and 20th centuries, Congress consistently conferred rule-making authority on administrative agencies. The Securities Exchange Act of 1934, a modern example of a Congressional mandate to the Securities and Exchange Commission (SEC), made unlawful "any manipulative or deceptive device or contrivance in contravention of such rules and regulations as the commission may prescribe as necessary or appropriate in the public interest or for the protection of investors." This statutory mandate served as the authority for SEC rule 10(b)-5, which placed restrictions on "insider securities trading." Likewise, the Occupational Safety and Health Act (OSHA), placing a general duty of employers "to prevent workplace hazards" and delegating authority to the Secretary of Labor to promulgate detailed "health and safety standards," is the basis for a variety of regulations such as maximum levels of exposure to toxic substances, prescribing guard rails, or requiring certain safety clothing.

The emergence of numerous regulatory administrative agencies at every level of government has been a phenomenon during the last half of the 20th century. The federal government has created the Internal Revenue Service, Interstate Commerce Commission, Environmental Protection Agency, Securities and Exchange Commission, National Labor Relations Board, and the Occupational Safety and Health Administration to name only a few. At the state government level, there are many counterpart agencies, numerous licensing boards, and a range of other agencies. At the county and city level there are zoning boards, school boards, and welfare agencies. All of these administrative agencies, whether called a board, authority, department, or commission, have powers to investigate, make rules, and supervise activities in various industries and in some limited geographical areas.

The growth of administrative agencies has been explained on two primary grounds. First, that certain specialized activities were best regulated by experts in the field of those activities and industries and, second, that legislatures and local governmental bodies were unwilling or ill-equipped to cope with the details of such regulation.

In the maze of regulatory programs involving either public or private engineering activities, a frequent source of friction is the lack of sophistication in the process, which is essential between the regulated person (e.g., engineers managing the project) and the regulators (e.g., government officials represented by politicians and bureaucrats). While politicians commonly are more concerned with their constituents' approval than with any technical details, engineers find that the indirect approach in which bureaucracy frequently functions is time consuming and tends to interfere with efficient completion of engineering projects.

The purpose of this chapter, which focuses on the policy, law, and activities of administrative agencies in general, seeks to furnish engineers a better understanding of the administrative process. The traditional analysis of administrative agencies concerns the limits placed on the powers and actions of the administrative agencies. These issues can be divided into (1) delegation of power, (2) exercise of quasi-legislative and quasi-judicial authority, (3) use of administrative procedures, and (4) intervention in and review of agency action by the judiciary.

Administrative Structure and Operation

While administrative agency action takes a variety of forms, agency action resulting in the issuance of rules or regulations with respect to certain conduct resembles the function of the legislature. Agencies acting in this quasi-legislative mode have the power to investigate situations coming to their attention and subsequently make rules or regulations based on their findings. For example, an appropriate agency may adopt a set of rules (regulations) establishing comprehensive environmental standards ranging from air quality and hazardous waste disposal to design and construction methods. Although there is no Constitutional requirement that Congress or state legislatures give any notice to or hear comments from their constituencies prior to adopting legislation, an administrative agency, by way of contrast, is generally required to give advance notice of proposed regulations and to listen to comments from persons who will be affected by such regulations. This does not require a traditional "hearing" and may be satisfied by merely receiving written comments as described earlier in this chapter.

In other instances, the agency performs functions similar to a court in performing some of the agency functions. In this style of function, referred to as adjudication, the agency action results in applying the agency's policies to some act already in place so an order and decision is issued against an individual or a business. The order or decision may require that a particular individual or business either take or cease certain action that might involve remedying an unfair labor practice by requiring reinstatement and back pay, modifying a particular building plan to eliminate pollution problems, or complying with safety or other standards. In this mode, the agency action resembles a court, and adjudication looks to the past. However, agencies need not wait until an injury has occurred before taking action since an agency may also exercise quasi-executive functions in carrying out investigations and gathering information. Agencies are given broad powers: (1) to direct a business to file annual or special reports and to answer specific questions in writing, (2) to obtain access to the files of a business, and (3) to subpoena witnesses and documents.

While early cases challenged agency action as violating the Constitutional separation of powers by combining both legislative and judicial functions, the modern trend of decisions is usually unsuccessful. The current thinking is that administrative agencies blend the separate powers to govern a segment of the population or business world in the public interest. The courts rationalize these delegations of authority as legally acceptable blendings of separate functions on the basis of being "quasi" rather than having full use of executive, legislative, or judicial power. The "quasi" designation uses statutory guidelines or limitations on the exercise of the delegated power, and this technique avoids the unconstitutional delegation condemned in *A.L.A. Schecter Poultry Corp. v. United States*, 290 U.S. 495 (1935), in which Justice Benjamin Cardozo wrote,

> [The power granted] . . . was not canalized within banks that keep it from overflowing. It is unconfined and vagrant . . . . [providing in effect] a roving commission to inquire into evils and upon discovery correct them.

By contrast, in a later environmental case, the Supreme Court approved the delegation of authority to the Environmental Protection Agency (EPA) to establish under the Clean Air Act national primary and secondary ambient air quality standards with guidelines

> . . . using latest scientific knowledge, to establish nationwide air-quality standards for each pollutant having an adverse effect upon public health or welfare. *South Terminal Corp. v. EPA*, 501 F.2d 646 (1947).

Some critics of administrative agencies contend that courts could carry out better the functions that agencies are directed to perform. But, in assessing that criticism, it is useful to consider the traditional nature of courts to see the magnitude of changes that would be required for courts to replace agencies. The principal function of courts is to hold trials. Traditional court procedure is designed to obtain all pertinent information regarding some past event, and, based on this information, the court makes a judgment. This is the court procedure expertise. Only rarely, and then in the nature of an equity action, will courts concern themselves with things that have not yet occurred. A court gives neither advisory opinions nor decides matters not in dispute, so it will not decide a matter that is moot, meaning there is no real and present issue. Action that formulates policy that will apply in the future to the general public falls within the ambit of rule making or legislating.

Federal Administrative Procedures Act

Traditionally, administrative agencies varied considerably in their organizational structure and in their rule-making and decision procedures. As the result of mounting criticism of this maze of divergent administrative style of action, President Franklin D. Roosevelt in 1939 appointed a committee to review the federal administrative process. After much debate and compromise, the Federal Administrative Procedure Act (FAPA) was adopted in 1946. The major effect of the FAPA satisfied the political demand for reform, to improve and strengthen the administrative process by regularizing certain procedures, to standardize the requirements for notice and hearing, to provide minimum standards of due process, and to set a criteria for judicial review. Generally, the states adopted similar legislation covering state and local administrative agencies after a Model State Administrative Procedure Act was promulgated in 1961.

14.2 POWER TO LEGISLATE: RULE MAKING

Legislative Function: Rule Making and Regulations.

The issuance of administrative regulations in the rule-making process is a major factor in administrative practice. The volume of administrative legislation is staggering. At the federal level alone the volume published annually in the *Federal Register* (which represents current additions and amendments) consistently exceeds 50,000 pages. To check all existing current federal regulations requires the review of over 65,000 pages in approximately 130 volumes of the *Code of Federal Regulations*. In considering the regulatory process, the policy makers must select between rule making and adjudication. The rule-making process has several advantages over the case-by-case process of adjudication, since it can quickly put all affected parties on notice of anticipated changes in regulatory policy with the regulatory program being drafted in a coordinated pattern rather than developed by isolated litigated cases. At the same time, it gives the parties an opportunity for limited participation in the agency policy by means of written comments before the regulation becomes finalized.

A centerpiece of the FAPA is the publication of proposed rules with an invitation for interested persons to make written comments. Under section 551(4) of the FAPA, a rule is defined as

> the whole or a part of an agency statement of general or particular application and future effect designed to implement, interpret, or prescribe law or policy.

This formulation carries out the usual concept of "general future application," which is a characteristic feature of legislation. The somewhat confusing reference to "particular applicability" results from an historical anomaly to preserve the traditional understanding that establishing rates and tariffs would be regarded as rule making rather than adjudication.

This broad definition of rule making has been weakened by three exceptions to the rule making procedures. First, certain functional rule making under Section 553(a) such as "a military or foreign affairs function . . . or agency management . . . was exempt from the prescribed procedures." Second, action with respect to "interpretative rules and general statements of policy . . . ." was excused from the normal procedures. Third, the language of Section 553(b)(3)(B) excused the agency from the notice procedures when the agency makes a finding of good cause that public proceedings "were impracticable, unnecessary, or contrary to the public interest."

Rule Making: Notice and Comment Procedures

When the rule-making procedures of Section 553 of the FAPA apply, it requires advance general notice of proposed rule making to be published in the *Federal Register* (official government document published daily) not less than 30 days before the effective date of the rule. This general publication constitutes legal notice to all affected parties and provides a technique for the decision maker to obtain information concerning the effect the proposed rule will have when implemented. On the other hand, this procedure provides the parties who oppose agency action with a voice in the final rule but with less opportunity to be heard than in a court-style adjudication. The procedure does ensure that the agency will be exposed to the factual, legal, and policy analysis of any party who cares to prepare and submit written comments. The court in *Automotive Parts & Accessories Association v. Boyd*, 407 F.2d 338 (D.C. Cir. 1968) indicated that the agency statement of basis and purpose for the rule would enable the reviewing court "to see what major issues of policy were ventilated by the informal proceedings and why the agency reacted to them as it did."

While the FAPA provides a minimal procedural standard, the enabling legislation establishing an agency or delegating to the agency the administration of a particular statute may set up a higher and more rigid standard for rule-making than that required by the FAPA.

Nonlegislative: Policy Statements

A general statement of policy, which is exempt from the notice requirements of rule making, is merely a disclosure to the public for keeping the public advised of the agency's thinking in the area of its responsibility. *Pacific Gas and Electric Co. v. FPC*, 506 F.2d 33, 38-39

(D.C. Cir. 1974) reflects the court's attitude concerning the effect and weight that should be accorded "policy statements" that are not based on the use of the public notice procedures. In that case, Judge George E. MacKinnon expressed the view that:

> A general statement of policy . . . is not finally determinative of the issues and rights to which it is addressed. The agency cannot apply or rely upon a general statement of policy as law because such a general statement only announces what the agency seeks to establish as policy. A policy statement announces the agency's tentative intentions for the future. When the agency applies the policy in a particular situation, it must be prepared to support the policy just as if the policy statement had never been issued. An agency cannot escape its responsibility to present evidence and reasoning supporting its substantive rules by announcing binding precedent in the form of a general statement of policy.

Nonlegislative: Interpretative Rules

An interpretative rule, which is another controversial area of agency action also exempt from the notice requirements of rule making, shows how the agency intends to apply the law and frequently is the agency's opinion on the meaning of the statute or one of its terms. The distinction between the legislative rules of an agency and the nonlegislative "interpretative rules" is that the former are binding on the court as an extension of legislative power while the latter have only the effect that the court, in its discretion, decides to afford them.

This exception for "interpretative rules" in FAPA Section 553, eliminating the notice requirements of the rule-making procedure, has been particularly troublesome since neither the Act nor its legislative history defines the meaning of "interpretative rules." While some courts have sought to distinguish legislative rules from interpretative rules by referring to the former by the term "substantive," this simple labeling dichotomy has the disadvantage of improperly suggesting that interpretative rules are limited to mere "procedural" matters. This shorthand analysis provides no real light since the use of the term "substantive" adds little to the establishment of a useful criteria of relevant factors.

However, the distinction is critical because agencies frequently adopt interpretative rules involving the determinative issue. The court in *Gibson Wine Co. v. Snyder*, 194 F.2d 329, 331 (D.C. Cir. 1952) suggests that interpretative rules are "statements as to what the administrative officer thinks the statute or regulation means," while legislative rules "create law, usually implementing an existing law." So, to say that a rule that construes a statutory term would be an interpretative rule when it represents an exercise of delegated lawmaking authority merely begs the question. In *Pharmaceutical Manufacturers Association v. Finch*, 307 F. Supp. 858 (1970), which appears at the end of this chapter, a qualitative test is used to inquire whether the so-called interpretative rules in question are sufficiently important and controversial to merit the safeguards of notice and comment procedures. Some courts side step the issue by holding that interpretative rules are not controlling but do constitute a body of experience and informed judgment to which the courts and litigants may properly resort for guidance. In effect, while not holding interpretative rules invalid because the agency failed to observe the public notice procedures, courts would limit their review of the interpretative rule to the "arbitrary and capricious" standard placing a heavy burden on the party challenging the interpretative rule.

14.3 POWER TO ADJUDICATE: ORDERS

Pleadings and Notice

Although procedures for formal adjudication by agencies differ depending on the regulatory subject matter, the FAPA establishes a basic minimum standard for administrative adjudications. Section 551(7) of the FAPA refers to "adjudication" as agency action that produces "an order." In turn, "an order" is defined as a final disposition of a matter other than rule making but including licensing. Generally speaking, adjudications are the procedures by which orders are issued after the presentation of evidence by the parties in a hearing (i.e., quasi-judicial adjudications) resembling court trial procedure. The FAPA establishes procedural requirements for adjudications that are "required by statute to be determined on the record [meaning that all the evidence is considered] after an opportunity for an agency hearing."

The jurisprudential foundation for adjudicatory notice announced by the Supreme Court in *Morgan v. United States*, 58 S. Ct. 473, 476 (1938) provides that "[t]hose who are brought into contest with the Government in a quasi-judicial proceeding aimed at the control of their activities are entitled to be fairly advised of what the Government proposes and to be heard upon its proposals before it issues its final command."

With respect to persons entitled to notice of an agency hearing, FAPA Section 554(b) codifies the criteria ". . . [that the parties] shall be timely informed of the time and place of the hearing, the legal authority that the agency is relying on," and "the matters of fact and law asserted." The judicial gloss of that section is described as founded on a policy of fairness, and these requisites are satisfied if there be a plain statement of the things claimed so that the parties may be put on their defense. *American Newspaper Publishers Association v. NLRB*, 193 F.2d 782, 800 (7th Cir. 1951).

Right to Hearing

The counterpart of a court trial is an agency hearing. The principal differences from the traditional court trial are that a hearing officer rather than a judge presides, and there is usually much less formality in the receipt and consideration of evidence. Since administrative hearings do not have juries, the rules of evidence to present confusion or prejudice to a juror do not apply. Generally, after the hearing, the hearing officer (in the case of the FAPA (an independent administrative law judge) who conducted the hearing prepares a preliminary decision, which includes findings of facts, conclusions of law, and recommended relief, which decision in turn is submitted to the agency for appropriate final action.

The purpose of an adjudicative hearing is to determine issues based on events which have already occurred. By way of illustration, a complainant (employee or union) files an unfair labor practice charge with the National Labor Relations Board (NLRB), and, in turn, the NLRB after investigation issues a complaint appointing a trial examiner to conduct a hearing. The hearing, a trial-type procedure, attempts to elicit the facts asserted in the complaint and defense. Witnesses are called, sworn, examined, and cross-examined; documentary evidence is introduced. Hearsay may be allowed, since the court rules of evidence are not strictly applied. The findings, conclusions and recommended relief as determined by the hearing officer are submitted to the NLRB. The NLRB is not bound by the hearing officer's recommendation, and it may affirm or modify the findings and remedial order. Depending upon the nature of the case and on other circumstances, an aggrieved party may appeal to a U.S. District Court or a U.S. Court of Appeals. This described hearing procedure is similar to that followed by nearly all federal administrative agencies and some state agencies. The principal difference at state and local government levels is that the hearing takes place before the entire agency rather than a hearing officer, and the appeal procedure will go through the state court system.

14.4 AVAILABILITY AND TIMING OF JUDICIAL REVIEW

Timing of Judicial Review

Judicial doctrine requires that the aggrieved party exhaust all administrative relief available including all administrative remedies and appeals before seeking judicial relief. The reviewing court frequently expresses the doctrine as a matter of jurisdiction in which the court declines jurisdiction over the issues until all administrative avenues have been exhausted. In other cases, the court expresses the view that under a doctrine referred to as "ripeness for review" the agency must be given the opportunity to monitor its own actions and decisions and to provide the court with the rationale and expertise of the agency in its final agency decision.

In an early case, a company, when served with an NLRB complaint alleging unfair labor practices sought judicial relief to enjoin the NLRB administrative proceeding. The company unsuccessfully urged lack of agency jurisdiction alleging that "the company's products were not sold in interstate commerce." The Court concluded that Congress intended that the NLRB exercise exclusive power to make the initial determination of whether or not the employment situation "affects interstate commerce." In response to the company's argument that it was being put to an unjustified expense of a trial, the Court held that "this expense was not irreparable harm . . . [adding that] . . . lawsuits often turn out to be groundless. But no way has been discovered of relieving defendant from the necessity of a trial to establish that fact."

In *McKart v. United States*, 89 S. Ct. 1657 (1969), the Court summarized the policy underlying the exhaustion doctrine as (1) permitting the statutory scheme to function by allowing the agency to apply its expertise and to exercise the discretion granted to the agency by the legislature, (2) providing efficiency by permitting the administrative process to proceed uninterrupted subject to judicial review at the conclusion of agency action, (3) protecting the autonomous nature of the agency that is not part of the judiciary, and (4) avoiding interim action to hinder the agency in its assembling and analyzing of facts and the preparation of the agency's basis for action.

In addition to the necessity that all administrative remedies must be exhausted, only "final" agency action is subject to judicial review. The requirement of finality permits the agency to reverse or modify its decision after further deliberation.

Administrative Discretion

Section 702 of the FAPA affords standing for judicial review to a "person suffering a legal wrong . . . or [being] adversely affected or aggrieved by agency action within the meaning of a relevant statute." Usually, the agency's organic statute contains a statutory provision setting forth the means by which a particular agency is to be reviewed by a court. Generally, specific statutory provisions give a party the right to petition the federal district court or the federal court of appeals to set aside the agency order or challenge the legality of a rule, regulation, or order. The system for judicial review of administrative action undertakes to balance the presumption of administrative expertise against the necessity of providing the interested party with regularity of legal authority, due process, fairness, and Constitutional rights. In establishing the scope and depth of judicial review, some recognition is given that, by their nature, administrative decisions are incapable of standing up to a highly critical judicial attitude. Accordingly, in defining the criteria, reviewing courts suggest that unless some risk of error is tolerated, a large percentage of agency action and the decision would fail to pass judicial scrutiny. For this reason, the administrative process has developed the "substantial evidence" and the "arbitrary or capricious" standards. These standards tell the court to monitor the agency for probability of correctness and builds into the administrative program a tolerance for error.

Scope of Review

Substantial evidence standard. For judicial review of factual findings, "substantial evidence" binds the court to the agency's conclusion unless the court finds the conclusion to be unreasonable. This standard of review conveys to the court that it need not delve so deeply into the agency's judgment as to assure that the conclusion is either the only acceptable finding or absolutely correct. The substantial evidence test establishes that the court's review be of a nature to assure that there is a relatively high probability concerning the findings made by the agency. To meet this test, the Court's opinion in *Consolidated Edison Co. v. NLRB*, 59 S. Ct. 206, 216 (1938), required that "there is such probative evidence as a reasonable mind might accept as adequate to support a conclusion." In a later case, the Court, in *Universal Camera Corp. v. NLRB*, 71 S. Ct. 456, 464 (1951), restated the standard as

> The evidence must be sufficient to support the conclusion of a reasonable person after considering the evidentiary record as a whole, not just that which is consistent with the agency's findings.

This criteria is less rigorous than traditional court standards which are variously described as the weight or the preponderance of the evidence tests. See, *Consolo v. FMC*, 86 S. Ct. 1018, 1026 (1966). Scholarly writers concede that the substantial evidence test still affords considerable deference to agency findings, and, where several findings are reasonable, the findings selected by the agency will be affirmed based on their reasonableness rather than the court's preference for another finding.

Arbitrary and capricious standard. Professors Ernest Gellhorn and Barry B. Boyer in their book, *Administrative Law and Process*, published in 1981 by West Publishing Company, suggest that reasonableness connotes the operation of a reasoning mind, and, hence, unreasonableness is a result that cannot be the product of a reasoning mind. This test compels an affirmative conclusion that the decision could not be the product of a valid reasoning process and beyond the boundaries of sound judgment. It is only when the regulation fails to meet this standard that the agency action will be declared invalid as arbitrary and capricious.

Since 1971, courts have looked to the Supreme Court's decision in *Citizen's to Preserve Overton Park v. Volpe*, 91 S. Ct. 814 (1971) for guidance in the application of the arbitrariness standard. *Overton Park* involved an informal adjudication by the Secretary of Transportation to free federal funds for a construction of a highway through Overton Park in Tennessee. Under the Department of Transportation Act and the Federal-Aid Highway Act, the secretary could not authorize the use of federal funds to construct highways through public parks "if a feasible and prudent alternative route existed." The plaintiffs contended that the secretary violated these statutes by authorizing a six-lane highway through the park. On review, the court after finding absence of any formal hearing requirement eliminated the "substantial evidence" review but concluded that "the reviewing court is required to engage in a substantial inquiry" and the presumption of regularity "is not a shield to [the secretary's] action from a thorough, probing, in depth review." To make a finding of nonarbitrariness, the Court concluded,

> that there must be consideration of whether the decision was based on a consideration of the relevant factors and whether there has been a clear error of judgment....Although the inquiry into the facts is to be searching and careful, the ultimate standard of review is a narrow one."

In *Motor Vehicle Manufacturers Association of the United States Inc. v. State Farm Mutual Automobile Insurance Co.*, 103 S. Ct. 2856 (1983), the Court found the action of the Secretary of Transportation to be arbitrary in rescinding the passive restraint (air bag) requirement. This case involved a long regulatory evolution toward a mandatory passive restraint system involving seat belts and then air bags. After a period of comments and public hearings, the secretary issued an order rescinding the program. The Court held that "[b]y failing to analyze the continuous use of seat belts in its own right, the agency failed to offer the rational connection between facts and judgment required to pass muster under the arbitrary and capricious standard."

Burden of proof. With respect to the question of the burden of nonpersuasion, Section 556(d) of the FAPA provides that "[e]xcept as otherwise provided by statute, the proponent of a rule or order has the burden of proof." Despite this language, which generally would make the government regulators the proponent, the courts have not been consistent in applying this provision, and the concept of determining which party is the proponent frequently presents a tricky and elusive question. For example, in *United States Steel Corp. v. Train*, 556 F.2d 822 (7th Cir. 1977), the court held that United States Steel, an applicant in a National Pollutant Discharge Elimination System (NPDES) permit proceeding, was the proponent since the steel company sought permission to discharge water not otherwise allowed under the Clean Water Act. In *Foster v. Seaton*, 271 F.2d 836 (D. C. Cir. 1959), the court, in reviewing Foster's claim of ownership in government lands based on the assertion of discovery of valuable minerals, without a clear explanation and possibly confusing the burden of going forward and the burden of persuasion, held that the claimant, as "proponent," bears the burden of proof rather than the government, which issued an order denying the claim. Finally, in *Environmental Defense Fund v. EPA*, 548 F.2d 998 (D. C. Cir. 1976), that challenged an order banning the use of certain pesticides, the court upheld EPA regulations stating that "EPA, the proponent of the suspension, has the burden of going forward to present an affirmative case for suspension but that the ultimate burden of persuasion rests with the proponent of the registration [i.e., the pesticide manufacturer]."

14.5 EXCLUSIVITY, PREJUDICE, AND ETHICS

Exclusivity of the record is recognized as a rudimentary administrative law principle. This rule, requiring the decision maker and the reviewing court to look only to the publicly produced evidence, bars *ex parte* evidence in all agency adjudicatory hearings. The evils of secret communications or the insidious effect of private pressures by external influences on the decision maker are obvious.

The dissenting opinion of Justice William O. Douglas in the *Morton case*, at the end of chapter 2, suggests instances where a company (or even a whole industry) may exert subtle pressure on an administrative agency to "influence" and even "capture a [target] regulatory agency which represents virtually the controlling law to the regulated industry" of which the company is a member. This implication is the thrust of *Sangamon Valley Television Corp. v. United States*, 269 F.2d 221 (D.C. Cir. 1959), where the court set aside the Federal Communications Commission (FCC) decision to shift a television channel from one city to another because of improper advances by one of the applicants to the FCC commissioners. While the license application was pending, the president of one of the applicants had a lunch separately with each of the commissioners, spoke privately with each, and gave each commissioner a Christmas turkey 2 years in a row. In condemning this conduct, the court said ". . . attempts to influence any member of the Commission . . . except by the recognized and public processes go to the very core of the Commission's quasi-judicial powers . . . ." Unfortunately, administrative officials and staff members not only are subject to pressures from outside but also are subject to similar influence from within the government since few administrators or staff members are able to resist calls from the White House, the governor's office, or Capitol Hill.

Closely related in terms of influence peddling is the so-called "revolving door" problem. This is the outgrowth of the common practice for persons to move to government service from private employment and then return later to private employment from their government service. In response to disclosures of wide-spread abuses, Congress enacted the Ethics in Government Act (1978). This Act permanently forbids a former official from representing private parties in matters in which the official directly participated during the period of government service and subjects the former official to a 1 year cooling-off period on personal advocacy before the official's former agency. In addition, the Act requires the filing of detailed financial reports by administrative officials as well as by legislative, executive, and judicial personnel.

Opponents of these restrictions argue that such strict conflict of interest rules discourage talented individuals from accepting government service because they are unwilling to jeopardize post-government employment opportunities. However, the breakdown of self-discipline and the failure to comply with accepted ethical conduct with respect to conflicts of interest made these statutory proscriptions necessary. In an effort to capitalize on "government contacts," individuals have sought to take advantage of private access to administrative decision makers by maximizing their private interests to the detriment of the greater public interests.

It is commonplace for engineers to find themselves in this highly charged arena. Engineering skills play a vital role in making critical decisions concerning design, reliability, and usefulness of products in both private and the vast government complex. In turn, it is essential for supplying and service companies' engineering employees and consultants to contact, assist, and provide information in private and public bid proposals and contract negotiations. In this context and in the effort to achieve personal success, engineers find themselves with compromising temptations testing their ethical mettle. Ethical codes provide guidance for balancing short-run individual gains with the broader long-run professional goals. While the codes of some engineering societies fail to assign a clear positive responsibility for the general welfare to their members, a number of major codes do make a strong requirement that engineers give a high priority in their professional objectives to "conduct which benefits the public." This finds expression in the four "Fundamental Principles" of the Engineers' Council for Professional Development (ECPD) Code of Ethics stating that "Engineers uphold and advance the integrity, honor and dignity of the engineering profession by using their knowledge and skill for the enhancement of human welfare."

O'KEEFFE v. SMITH, HINCHMAN & GRYLLS ASSOCIATES
380 U.S. 359, 85 S. Ct. 1012 (1965)

[Ecker was employed in Seoul, Korea, as an assistant administrative officer for Smith, Hinchman & Grylls Associates, Inc., an engineering management concern working under contracts with the U.S. and South Korean governments. His duties were restricted to Seoul where he was responsible for personnel in the stenographic and clerical departments. He was subject to call at the job site at any time, but the usual work week was 44 hours, and employees were accustomed to travel far from the job site on weekends and holidays for recreational purposes. Ecker did not live at the job site; he was given an allowance to live in Seoul. On his Memorial Day weekend he went to a lake 30 miles east of Seoul where a friend of his (not a coemployee) had a house. Ecker intended to spend the holiday there with his friend and another visitor. Their Saturday afternoon project was to fill in the beach in front of the house with sand, but none was readily available. To obtain it, the three crossed the lake in a small aluminum boat to a sandy part of the shore. There, they filled the boat with a load of sand, intending to transport it back to the house. The return trip, however, put Archimedes' Principle to the test; in the middle of the lake the boat capsized and sank. Two of the three men drowned, including Ecker.]

Based upon the . . . stipulated facts, the Deputy Commissioner of the Bureau of Employees' Compensation, United States Department of

Labor, [petitioner] determined "that the accident and the subsequent death of the decedent arose out of and in the course of employment." He therefore awarded death benefits . . . [T]he employer . . . then brought this action to set aside . . . the enforcement of this compensation award. The District Court affirmed the compensation award . . . [T]he Court of Appeals . . . summarily reversed . . . . The judgment of the Court of Appeals is reversed.

In cases decided both before and after the passage of the Administrative Procedures Act . . . the Court has held that the [statute limits] . . . the scope of judicial review of the Deputy Commissioner's determination that a "particular injury arose out of and in the course of employment." It matters not that the basic facts from which the Deputy Commissioner draws this inference are undisputed rather than controverted . . . . It is likewise immaterial that the facts permit the drawing of diverse inferences. The Deputy Commissioner alone is charged with the duty of initially selecting the inference which seems most reasonable and his choice, if otherwise sustainable, may not be disturbed by a reviewing court. Moreover, the fact that the inference of the type here made by the Deputy Commissioner involves an application of a broad statutory term or phrase to a specific set of facts gives rise to no greater scope of judicial review . . . . "

The rule of judicial review therefore emerged that the inferences drawn by the Deputy Commissioner are to be accepted unless they are irrational or "unsupported by substantial evidence on the record . . . as a whole." *O'Leary v. Brown-Pacific-Maxon, Inc.*, 71 S. Ct. 470 (1951).

The Court in *Brown-Pacific-Maxon* drew the line only at cases where an employee had become "so thoroughly disconnected from the service of his employer that it would be entirely unreasonable to say that injuries suffered by him arose out of and in the course of his employment." This standard is in accord with the humanitarian nature of the Act as exemplified by the statutory command that "in any proceeding for the enforcement of a claim for compensation under this chapter it shall be presumed in the absence of substantial evidence to the contrary . . . that the claim comes within the provisions of this chapter." While this Court may not have reached the same conclusion as the Deputy Commissioner, it cannot be said that his holding—that the decedent's death, in a zone of danger, arose out of and in the course of his employment—is irrational or without substantial evidence on the record as a whole.

Justice Harlan Dissenting. I see no meaningful interpretation of the statute which will support this result except a rule that any decision made by a Deputy Commissioner must be upheld.

Whether the injury is compensable should depend to some degree on the cause of the injury as well as the time of day, location, and momentary activity of the employee at the time of the accident. I would distinguish between a case in which Ecker smashed his hand in a filing cabinet while at the office and one in which he tripped over a pebble while off on a weekend hike. In the first case Ecker's injury would have arisen out of and in the course of his employment, whereas the statute would not apply to the second case unless the injury were traceable to some special danger peculiar to the employment, which was clearly not the case. Thus, if while off on that same weekend hike Ecker stepped on a mine left over from the Korean conflict, a different result would follow.

Justice Douglas Dissenting. Reviewing courts must be influenced by a feeling that they are not to abdicate the conventional judicial function . . . . The [agency's] findings are entitled to respect; but they must nonetheless be set aside when the record before a Court of Appeals clearly precludes the [agency's] decision from being justified by a fair estimate of the worth of the testimony of witnesses or its informed judgment on matters within its special competence or both.

SOGLIN v. KAUFFMAN
418 F.2d 163 (7th Cir. 1969)

[Plaintiffs, students at the University of Wisconsin and members of Students for a Democratic Society, were charged by defendant University officials with "misconduct" consisting of physical obstruction of university buildings in order to prevent representatives of the Dow Chemical Company from conducting job interviews. Defendants instituted disciplinary proceedings to expel plaintiffs, who brought this action contending that "misconduct" is an unduly vague standard and that disciplinary proceedings for "misconduct" therefore violate the due process clause of the 14th Amendment. The District Court held that the university's action based on "misconduct" was unconstitutional.]

Defendants argue that "misconduct" represents the inherent power of the University to discipline students and that this power may be exercised without the necessity of relying on a specific rule of conduct. This rationale would justify the ad hoc imposition of discipline without reference to any pre-existing standards of conduct so long as the objectionable behavior could be called misconduct at some later date. No one disputes the power of the University to protect itself by means of disciplinary action against disruptive students. Power to punish and the rules defining the exercise of that power are not, however, identical.

Power alone does not supply the standards needed to determine its application to types of behavior or specific instances of "misconduct". As Professor Fuller has observed: "The first desideratum of a system for subjecting human conduct to the governance of rule is an obvious one: there must be rules." . . . The use of "misconduct" as a standard in imposing the penalties threatened here must . . . fall for vagueness. The inadequacy of the rule is apparent on its face. It contains no clues which would assist a student, an administrator or a reviewing judge in determining whether conduct not transgressing statutes is susceptible to punishment by the University as "misconduct."

Pursuant to appropriate rule or regulation, the University has the power to maintain order by suspension or expulsion of disruptive students. Requiring that such sanctions be administered in accord with pre-existing rules does not place an unwarranted burden upon University administration. We do not require University Codes of Conduct to satisfy the same rigorous standards as criminal statutes. We only hold that expulsion and prolonged suspension may not be imposed on students by a University simply on the basis of allegations of "misconduct" without reference to any pre-existing rule which supplies an adequate guide. The possibility of the sweeping application of the standard of "misconduct" to protected activities does not comport with the guarantees of the 1st and 14th Amendments. The desired end must be more narrowly achieved.

Judgment of the District Court holding that disciplinary proceedings based on a standard of "misconduct" were unconstitutional, is affirmed.

DISCUSSION QUESTIONS

1. Describe and distinguish the rule-making and adjudicative functions of an administrative agency.

2. Discuss the scope of judicial review of (1) regulations and (2) orders.

3. What is the rationale for the establishment of administrative agencies?

4. What are the underlying reasons for providing advance notice of proposed rule making (regulations)?

5. On what grounds did the court invalidate the university disciplinary action of the students in the *Soglin case*?

6. Does the scope of judicial review in the *O'Keeffe case* ensure that the administrative action will be wise and correct?

7. A young graduate, X, after receiving a master's degree from a prestigious engineering college, was employed by the U.S. Department of Defense in the weapons procurement division. In that assignment, X served as a member of a project team and routinely attended regular planning sessions with senior personnel in designing a new missile system. After X had been in the capacity for about 3 years, he was approached by a large corporation that is heavily involved in the defense industry. The corporation made him an attractive offer to join their weapons systems division. Not only was the salary quite attractive, but the perks included assistance by the corporation in completing his Ph.D. work. His immediate supervisor explained that the corporation hoped to be a successful bidder on the new missile system and especially on related hardware to be developed in the future. It was described as "you can hit the ground running since we are looking to put in a bid proposal in the area you have been working." Discuss the legal and ethical aspects of X's new work (1) assuming that X has photocopied personal copies of various memorandum relating to the defense objectives, (2) assuming no reproduced copies but only that X will use the general information with which he is familiar, and (3) assume that X does not directly participate as a member of the negotiating committee with the Department of Defense but serves in a support capacity.

8. A state statute permits a professional engineering society (association) to adopt a code of ethics and to police the conduct of members of the profession with respect to their compliance with the code. The association adopted a code of ethics that provided, among other things, that "engineers shall not engage in unfair competition" and "engineers shall conduct the professional relationships in an ethical manner." Engineer X has been cited "for violating the professional code of ethics" based on a charge filed by engineer Y alleging that X ran an advertisement in the *Yellow Pages* stating "BEST ENGINEERING—LOWEST RATES." Assuming that X authorized this advertisement, what are the legal defenses that X may interpose?

15

FACT FINDING
Use of Expert Evidence

15.1 THEORY OF FACT FINDING

General

Justice Benjamin Cardozo, a Supreme Court Justice and recognized legal scholar, who helped shape many contours of legal jurisprudence, described the relationship between facts and legal principles as

> More and more, we lawyers are awakening to a perception of the truth that what divides and distracts us in the solution of a legal problem is not so much uncertainty about the law as uncertainty about the facts—the facts which generate the law. Let the facts be known as they are, and the law will sprout from the seed and turn its branches toward the light.

In many disputes and especially in matters involving engineering or other scientific and technical questions, the development of an understanding of the facts is a critical issue. This is especially true when the triers of fact (whether the fact finder is a juror, court, or administrative agency) are called upon to determine a key factual issue that is beyond the scope of their ordinary experience.

Historical Development of Trials

Evidence is offered for the purpose of proving or disproving certain facts in controversy. While modern procedure relies heavily on evidence produced by means of witnesses, the use of witnesses as sources of factual evidence in a trial did not develop at Common Law until the use of a jury as a fact finder was firmly established.

Early trials were more mechanical than their modern counterparts. However, the jury trial gradually replaced the older forms of trial (i.e., the ancient species of trial by fire ordeal and water ordeal). Early juries were composed of neighbors either already acquainted with the facts under dispute or in a position to be capable of easily discovering the disputed facts. Since there was no settled practice for obtaining the facts by means of witnesses, it was not until the 16th century that Parliament compelled witnesses to attend and testify in the Common Law courts.

15.2   REQUIREMENT OF KNOWLEDGE BY WITNESS

Actual Observation

In the Common Law system, proof of facts was exacting in its insistence upon reliable sources of information. This rule was pervasively illustrated by the requirement that a witness who testified concerning a fact that could be perceived by the senses must have had an opportunity to observe and actually had observed the fact. Under this concept, if the testimony of a witness, either on its face or in its form, purports to be testimony of observed facts, but is in fact only repetition of statements of others, that testimony is inadmissible. The court would sustain an objection on the grounds that the witness lacked first-hand knowledge.

While the law is well established with respect to the necessity of first-hand observation, it is not so impractical as to insist upon preciseness of attention. Accordingly, when a witness uses expressions such as "I think" or "it is my impression," an objection to such testimony would not be sustained if it appears that the witness merely speaks from an inattentive observation. In *Ewing v. Russell*, 137 N.W.2d 892 (S.D. 1965), the court permitted testimony concerning the condition of a floor although the witness added that "she didn't pay any particular attention to the floor." These forms of expression refer to the quality of the witness' attention and not admissibility based on lack of first-hand knowledge.

Evolution of the Rule Against Opinions

Among the most important exclusionary rules in the law of evidence is the "opinion" rule. The requirement that witnesses must have personal knowledge is a very old rule, having its roots in medieval law. The rule demands that witnesses speak only "what they see and hear." Although the opinion rule is a practice developed by the English courts, it is now more rigidly enforced in the U.S. courts than in the English courts. In its modern form, this rule states that a witness testifying on the issues before the court can offer only "facts" personally and directly observed. The underlying rationale is that once the facts are known, a jury or other fact finder can be expected generally to draw the proper inference, to form an appropriate opinion or reach a correct conclusion.

In analysis, however, there is no conceivable statement, however specific, detailed, and "factual," that is not in some measure the product of inference and reflection as well as observation and memory. The difference between the statement, "He was driving on the left-hand side of the road," which is classified as fact, and "He was driving carelessly," which is called "opinion," represents merely a difference between a more concrete and specific form of descriptive statement and a less specific and concrete form. Thus, the difference between so-called "fact" and "opinion" is not a clear cut distinction between opposite absolutes but merely a variant in degree without a litmus test to mark the boundary.

In U.S. courts it now appears that some of the trial judges lean toward the principle espoused by Professor John Henry Wigmore, the leading scholar on the Law of Evidence, namely, that opinions of laymen should be rejected only when they are superfluous in the sense that they will be of no value to the jury. This test was applied in *Allen v. Matson Navigation Co.*, 255 F.2d 273, 278 (9th Cir. 1958), where a witness was allowed to testify that the floor was "slippery." However, the opinion rule essentially operates to prefer the more concrete description ("there was water spilled on the tile floor") to the less definite description. In turn, as the circle of facts comes closer to the hub of the "ultimate" issue, the courts are more reluctant to receive opinions or inferences. Accordingly, it is clear that evidentiary conclusions by a lay witness that the defendant was negligent or breached a contract are inadmissible.

The psychological prejudicial effect of the opinion "slippery" may be impossible to overcome, although an attempt is made to discredit that "empty opinion" through skillful cross-examination by inquiring, "What was the substance on the floor? How did you know? What was the composition of the floor: tile, wood, or concrete? Was the substance wet, dried, or hardened? Had the area been mopped, covered, or roped off?" Accordingly, with respect to matters within the scope of common knowledge or the shared experience, the thrust of the rule is that the testimony of witnesses must generally be confined to statements of actually observed facts gathered by the use of an individual's own senses.

15.3 OPINION RULE: EXPERT WITNESS EXCEPTION

While the opinion rule limits the ordinary layperson witness to matters actually within that person's factual knowledge, the rule has an important exception: It does not apply to expert witnesses. An expert witness can offer analysis and opinions beyond the mere presentation of facts. Expert testimony deals with matters beyond the knowledge and competence of the fact finder, and, for this reason, expert testimony is accepted in ways that ordinary testimony is not.

Underlying this exception is the fact that when the subject matter at issue is sufficiently beyond the knowledge of a layperson, expert testimony must be used to assist the jurors or fact finder in resolving the disputed issues. The use of an expert witness is especially common in engineering, technical, or scientific disputes dealing with construction and manufacturing questions. For example, a consulting engineer testified with respect to construction, safety, and the characteristics of terrazzo steps in *Associated Dry Goods Corp. v. Drake*, 394 F.2d 637 (8th Cir. 1968); an architect testified concerning the condition of a stair in *Nancy v. Lane*, 428 P.2d 722 (Ore. 1967); a mechanical engineer testified as to the suitability of the design of a snap-tie for supporting scaffolding in *Gates & Sons, Inc. v. Brock*, 199 So.2d 291 (Fla. 1967); and a civil engineer indicated the effect of certain land development on flooding in *Kuklinska v. Maplewood homes Inc.*, 146 N.E.2d 523 (Mass. 1957).

Forrest L. Tozer, in "Preparation and Use of Technical Evidence in Products Liability Cases," 16 *Defense L.J.* 669 (1967), succinctly poses the need for expert testimony:

> How can we hope to communicate sophisticated knowledge of technical facts to a jury of six retired mailmen and six bored housewives? How can we hope to make instant chemists or physicists out of the jurors? How can we hope to get across to a jury the intricacies of electronic production-control equipment, or structural principles on which a design is based, or the knowledge of bio-chemistry necessary to judge whether a particular drug could have caused the condition from which the plaintiff suffers?

Experts with First-Hand Knowledge

Often, experts are called as witnesses because of first-hand knowledge by the witness concerning the issues in dispute. Any expert who personally observed the pertinent facts may describe those observations and offer an opinion based on them. Examples are physicians who personally examined the claimants in personal injury cases or engineers who investigated and examined the physical facilities after a structural failure occurred. In *Fernholtz Machine Co. v. Wilson*, 5 P.2d 679 (Cal. App. 1931), the California court permitted a mechanical engineer, who was familiar with a particular type of machine and the kind of work it performed, to offer an opinion on whether the machine was completed in conformity with the controlling specifications.

It is the practice in some courts to require the expert to disclose on direct examination the underlying facts or data on which the opinion is based. However, other courts do not require the expert to state the underlying facts or data at the outset but leave that disclosure to cross-examination. Cal. Evid. Code Section 802 (West); Fed. Rules of Evid. Rule 705.

Manner of Acquisition of Facts

The means by which the expert acquires the information that is the basis of the expert opinion of the witness may involve questionable or improper means resulting in the exclusion of the expert's opinion. In *Agent Orange Product Liability Litigation*, 611 F. Supp. 1223 (E.D.N.Y. 1985), Judge Jack B. Weinstein found a hematologist's and a pathologist's expert opinions to be inadequate to overcome defendant's motion for summary judgment because of the manner in which the experts acquired their data. After the action had been filed for the benefit of Vietnam veterans, plaintiff's counsel asked those veterans to fill out a symptomatology checklist and to indicate any physical, mental, or emotional symptoms. Even though the information on which the opinion is based is of the "kind" that an examining physician would ordinarily require, the expert's opinions were excluded because of the manner in which information was obtained. The court found inadequate the plaintiff's evidence based on the doctor's expert opinions since a physician normally would not rely on this information without corroboration from medical records, physical examinations, and medical tests. A similar result occurred in *Swine Flu Immunization Products Liability Litigation*, 508 F. Supp. 897 (D. Colo. 1981), which involved the issue of the causal link between the swine flu vaccination and the Guillain-Barre Syndrome. The court held that "the technique employed in compiling data . . . [collected from hospitals by plaintiff's counsel] . . .

is not of the caliber used in the field of epidemiology and medical statistics and therefore could not be the basis of the [expert] opinion testimony." Again, the court's decision rests on criticism of the quality of raw facts, more precisely on "the means by which the raw facts were gathered," which was the basis of the expert opinion.

Ultimate Issue Determination

Prior to adoption of Federal Rule 704, the earlier prevailing rule excluded testimony as to the ultimate issue as "invading the province of the fact finder" or "usurping the function of the fact finder." Professor Wigmore rejected this earlier restrictive rule as a "bit of empty rhetoric" by pointing out correctly that the fact finder (whether juror, court, or administrative agency) could always choose to reject the witness' testimony and make a finding to the contrary.

However, implementation of Rule 704 permitting expert testimony on the ultimate issue presents practical problems. For example, in many negligence tort cases, the ultimate issue is whether the defendant's negligence was the "proximate cause" of the plaintiff's injury. It is common practice in situations, such as a building collapse, to finally ask, "Was the defendant negligent in making the design" and "Was that design the proximate cause of the building's failure?" This brings into focus the precise meaning of "proximate cause," which is defined as

> a moving or effective cause or fault which, in a natural and continuous sequence (unbroken by an efficient intervening cause) produces the (harm, accident, injury, collision, occurrence) which otherwise would not have occurred.

or

> the act or omission complained of must be such that a person using ordinary care would have foreseen the event, or some similar event, which might reasonably result therefrom.

Dean Page Keeton, a leading authority on the Law of Torts and tort litigation, describes the fact-finding process as performing an "evaluation determination," in which the fact finder applies the legal standard of foreseeability to the established facts. Thus, two points must be decided: first, the traditional question of proof of the alleged facts of negligence (i.e., what if conduct constituting fact A occurred) and, second, the evaluation application of the legal standards to those facts (i.e., the question of the legal relationship between the established conduct (fact A) and the injury). In effect, critics of the expanded role of

the expert witness argue that an engineer testifying to design or construction of an alleged defective device would require that the engineer
be given a law education concerning the policy and meaning of
"proximate cause" in order to testify on that ultimate issue.

Hypothetical Questions

The expert witness is in a unique position to render an opinion,
draw inferences, or make conclusions that the lay fact finder would not
be competent to evaluate by merely knowing the bare facts. The expert
is also qualified to explain or discuss technical, scientific, or other specialized knowledge bearing on the case, leaving it to the jurors or fact
finders to apply such knowledge to the facts.

A hypothetical question is a form of question in which counsel
states facts claimed or assumed to have been proven and then proposes
a question asking for an opinion of the expert based on those facts. The
traditional view is that an expert may state an opinion based upon
either first-hand knowledge or facts in the record at the time the
opinion is expressed. In addition, the opinion may be based partly on
first-hand knowledge and partly on facts in the record. If the expert's
opinion is to be based on facts in the record, the expert may obtain
such facts by virtue of having been present during the testimony or the
witness may be furnished a statement of facts in a hypothetical question. A hypothetical question, properly framed, should be sufficiently
specific and clearly stated so that the jurors or the fact finder will know
with certainty upon what state of assumed facts the expert opinion is
based. On cross-examination, the adversary may supply the expert with
omitted facts or ask the expert to assume other facts and then inquire
whether the expert's opinion would be modified.

The use of hypothetical questions to experts is an ingenious and
logical means of enabling the jurors or the fact finder to obtain the
expert's scientific knowledge concerning the facts of the case. Nevertheless, the use of the hypothetical question has been severely criticized in
the scholarly legal literature. First, if the examiner must recite all of
the relevant facts, the question becomes extremely wordy. Second, if
the examiner is permitted to select facts for the premise of the question, the hypothesis becomes one sided as per the old truism, "Ask the
wrong question and get the wrong answer," which translates in modern
computerese to "garbage in garbage out."

In a classic passage, Professor Wigmore called for the abolition of
the hypothetical question by stating,

> The hypothetical question, misused by the clumsy and abused by the
> clever, has in practice led to intolerable obstruction of truth. In the first

place, it has artificially clamped the mouth of the expert witness, so that his answer to a complex question may not express [the expert's] actual opinion on the actual case. This is because the question may be so built up and contrived by counsel as to represent only a partisan conclusion. In the second place, it has tended to mislead the jury as to the purport of the actual expert opinion. This is due to the same reason. In the third place, it has tended to confuse the jury, so that its employment becomes a mere waste of time and a futile obstruction. 2 Wigmore, *Evidence*, Section 686 (Chadbourn rev. 1979).

15.4 QUALIFICATIONS OF EXPERT WITNESS

Generally, an expert witness is qualified to offer an opinion if the court finds that because of certain skill, experience, education, or training the expert is better able to form an accurate opinion on the matter under consideration than an ordinary juror or fact finder. The expert's qualifications are determined relative to the particular subject on which the opinion is offered in evidence. Where the area of proposed testimony involves a field in which a license is required to practice, the court may consider the absence of a license as a condition bearing on the disqualification of the expert. The expert's qualifications are determined by the court as a preliminary matter and on appeal, the court's determination will not be reversed unless the determination was clearly in error.

In addition to the first-hand knowledge that the ordinary witness has concerning the issues in dispute, the expert has training, experience, or learning not possessed by the ordinary fact finder. This difference is a power to draw inferences from the facts that a juror or other fact finder would not be competent to draw. To warrant the use of expert testimony, then, two elements are essential. First, the subject of the inference must be so distinctively related to some science, profession, business, or occupation as to be beyond the understanding of the average layperson. *Manhattan Oil Co. v. Mosby*, 72 F.2d 840 (8th Cir. 1934). Second, the expert witness must have sufficient skill, knowledge or experience in that field or calling as to make it appear that the expert's opinion will probably aid the juror or fact finder in the search for truth. *Pennsylvania Threshermen Ins. Co. v. Messinger*, 29 A. 2d 653 (Md. 1943).

The knowledge in some fields may be derived from reading, in some from practice, or as is more commonly the case, from both. To be qualified, the expert may have acquired expertise by observation or experience and it is not necessary that the expert possess academic attainments or that the expert possess a professional degree. *Hoffman v.*

Chapter 15

Lindquist, 234 P.2d 34 (Cal. 1951); *Webb v. Olin Mathieson Corp.*, 342 P.2d 1094 (Utah 1959). In a case, an employee of defendant airline was permitted to testify as an expert witness concerning the failure of an aircraft horizontal stabilizer, although having had no scholastic standing, the witness had flown private planes, worked 11 months as an apprentice mechanic, and studied aerodynamics.

United States v. Amaral, (1973) sets out the traditional four basic requirements governing the admissibility of expert testimony:

(1) The subject matter must be beyond the common understanding of the average juror or must assist the juror in understanding the evidence;

(2) The expert must be sufficiently qualified so that his or her opinion or inference will aid the jury;

(3) The evidence about which the expert testifies must be scientifically reliable and generally accepted in the scientific community; and

(4) The probative value of the evidence must outweigh its prejudicial effect.

Federal Rule of Evidence 702 states a somewhat more lenient test and only requires (1) that the expert be qualified through skill, knowledge, or experience and (2) that the expert be able to help the trier of fact understand the evidence or determine a fact issue. Even if the evidence falls within the understanding of the average juror, an expert is still allowed to testify if that testimony adds to the jurors' understanding of the evidence.

15.5 LEARNED TREATISE AND THE EXPERT

A treatise basically expresses an author's (expert) views on a particular factual situation or branch of learning. However, the contents of a treatise generally are not admissible since the treatise, as evidence, constitutes out-of-court statements while the author is not subject to cross-examination. On the other hand, an expert may rely on an authoritative or specialized text as the basis for the opinion of the witness by testifying that the same scientific view is held by authors A and B. In that situation, the expert who already possesses professional experience and learning of the subject matter may be cross-examined as to the basis for the opinion expressed and asked if authors C and D did not hold opposite scientific views. The purpose of this cross-examination may be to demonstrate that the expert witness holds or expresses views that are contradicted by widely accepted authorities in the field.

15.6   EXPERT WITNESSES: ETHICAL CONSIDERATIONS

The escalation of litigation dealing with technical matters (e.g., environmental, medical, economic, and scientific issues) has spawned an explosion in the use of expert witnesses. An advertisement of one organization describes its services as including more than 12,500 experts available for litigation support in more than 4,000 categories. Over the years, the National Forensic Center (NFC), publisher of the bimonthly newsletter "The Expert and the Law" (Princeton, N.J.) evolved as a nation-wide representative of forensic scientists and reported that, at its national convention, its members expressed concern about the ethics and standards of performance of expert witnesses. NFC serves as a spearhead seeking to avoid a public image of the expert witness as a "prostitute," where the expert opinion is molded to fit the needs of whichever litigant is the highest bidder.

Some of the problems arise from the ignorance of the expert witness concerning the nature of the judicial system by failing to distinguish the role of the attorney from that of the expert witness. While the attorney properly assumes an adversary role many experts mistakenly view their role as simply an extension of the attorney in presenting the case.

A fundamental aspect of meeting the ethical standards required of the expert witness is an understanding of the dispute resolution process and the environment in which it operates. The expert witness must recognize the unique role of the expert as an instrument to clarify facts and issues in the search for justice and not as a "gun for hire."

Professionally, expert witnesses find themselves in the real world setting in which it is hard to turn down business. Any economic undertaking, whether it be an investment, a business or a professional practice faces the ultimate decision of whether it will be performed in a manner that will preserve its long term stability and usefulness, or in a manner that will maximize its short-term gains. Greed is certainly an element that is both a cause and effect of the latter approach.

A critical question is whether expert witnesses can or will establish and enforce a Code of Ethics. Failure to do so will leave the responsibility to the legislative and judicial branches to dictate the standards of performance for the expert witness. Already, Maryland, now followed by Kansas, established a legislative response to the problem by adopting a law which generally prohibits a party in a medical malpractice suit, when more than $5,000 in damages is sought, from presenting testimony from more than two experts in a designated specialty before an arbitration. Experts are further limited to those who devote a minimum of 80% of their annual professional time to practice in the area of their expertise. If more than 20% of the expert's time is spent on work that involves testimony in personal injury litigation, the expert does not qualify.

CITY OF SOUTH PORTLAND v. PINE STATE BY-PRODUCTS, INC.
306 A.2d 1 (Maine 1973)

Webber, J. Defendant corporation, a rendering plant engaged in converting fish, meat and poultry waste products into commercially salable materials, was permanently enjoined from causing or allowing offensive odors injurious or dangerous to the health, comfort or property of individuals or of the public to escape from the defendant's plant to such a degree that they are detectable . . . in the vicinity of its plant by a person of normal or average sensitivity to odors.

[T]he plaintiff, [the city in which] defendant's plant is located, filed its complaint charging that defendant disobeyed [the injunctive order and was in contempt. On trial, the court found for defendant and plaintiff appeals].

Appellant [plaintiff] . . . contends that reversible error was committed when an expert witness offered by defendant was permitted to express an opinion and advance a theory outside the realm of his training or experience. Dr. Amos Turk, admittedly a qualified expert witness with extensive practical and theoretical training and experience in the field of detection and control of industrial odors and with particular knowledge of the problems and control system in defendant's plant, gave a detailed explanation of that control system and reached certain conclusions with respect to its effectiveness. [That this expert opinion was persuasive to the court below was evidenced by the following findings:]

> I was much impressed by the testimony of Dr. Turk who was called as witness for the defendant. That he is an expert in the field of odor control is established by concession. It was his unqualified judgment that the odors complained of could not have come from the defendant's plant in the absence of any breakdown of the odor control equipment or the use of material which had putrefied to a degree it gave off odor beyond the control of the odor control system.
>
> There was no evidence whatsoever that the odor control equipment had broken down any time or that the plant had used unsuitable material in its processing. As a matter of fact there was evidence and I do find that

on some occasions when the complaints of odors were made the plant was not in operation. . . . In the present instant I am not satisfied from the evidence that this plant has been the source of any noxious odors on the dates on which complaints were registered.

The plaintiff had presented a number of witnesses, private citizens and South Portland police officers, who describe their detection of noxious odors on various dates which they concluded had emanated from defendant's plant. None of these witnesses, however, had entered the defendant's plant to observe whether the plant was operating or whether the control system was functioning. It is apparent that neither the defendant nor the justice below believed that these witnesses were knowingly and intentionally giving false testimony. It was the theory of the defendant, ultimately accepted by [the court below], that these witnesses did in fact detect noxious odors but were honestly mistaken as to their source. In this connection the findings state:

> Neither the sincerity nor the truthfulness nor the accuracy of any of the complainants is in doubt in my mind. I have come to the irresistible conclusion from the evidence that from time to time there were odors so noxious as to produce nausea in the complainants. I have no doubt as to the sincerity or truthfulness of the investigating officers. I do have serious doubts as to the method by which they arrived at their conclusions that the noxious odors emanated from this plant.
>
> I am satisfied, however, that though the officers who investigated and testified were most certainly in good faith when they attributed these specific odors which they detected to this plant, they could well have been in error in attributing the odors to this particular source.

In support of its theory that plaintiff's witnesses, though not untruthful, were honestly mistaken, the defendant elicited from Dr. Turk, over plaintiff's objection, testimony explanatory of the theory of "false alarm." In effect the witness stated that he and other experts engaged in the field of sensory evaluation of materials and detection problems have determined on the basis of their own experience and that of others that one may be predisposed to expect a certain odor from a certain source and as a result sometimes conclude that he detects an odor which does not exist or that an odor he detects emanates from an expected source when in fact it emanates from a different source. The objection to this evidence was grounded on the contention that the "false alarm" theory involved psychiatry and psychology, fields in which the witness had no training and professed no special competence. We discern no error in admitting the testimony. The discretionary ruling of the justice below as to the qualifications of an expert witness did not depend upon assigning labels to particular learning and experience of the witnesses. The witness demonstrated knowledge and competence in the area of sensory perception of odors which fully qualified him to furnish an opinion as to possibility of human error in detection and the reasons for it. The main issue, whether or not noxious odors emanated from defendant's plant, was decided on the basis of other evidence. The "false alarm" theory related only to whether the plaintiff's witnesses were honestly mistaken. In either event [the court below] was not disposed to accept their testimony as probative. It is understandable that [the court below] should not wish to have the public or the plaintiff's witnesses infer from a finding adverse to [plaintiff] that he had concluded that three witnesses had been other than truthful. Dr. Turks's theory did no more than to support his conviction that they had made an understandable human error in detection.

Appellant's second contention relates to the fact finder's alleged misuse of a view of defendant's premises. At the request of both parties [the court below] visited the area in the company of opposing counsel. When the hearing was resumed he spread upon the record a detailed report of the observations made at the scene. This included the detection of a a number of odors in the area and their apparent sources, as well as a negative observation as to any unpleasant odor emanating from defendant's plant.

Clearly it would have been error for the fact finder to find that ". . . because no offensive odors were emanating from defendant's plant at the time of the view, the same condition obtained on other occasions earlier in the summer . . . [A] view is not evidence and is taken only to assist the fact finder in better understanding the evidence otherwise produced.

On the basis of credible evidence [the court below] was persuaded that on the dates of the complaints the odors could not and did not originate in defendant's plants. He also had evidence from witnesses to the presence and sources of other noxious odors in the area, particularly odors emanating from the Stauffer Chemical Co. plant close by the defendant's premises. What he saw and smelled in the course of the view enabled him to better understand the testimony and, above all, comprehend how it was possible for a number if truthful witnesses to fall into the same error as to the source of the offensive odor they detected. It may be noted, for example, that none of these witnesses was aware that Stauffer Chemical Co. was also at times the source of an industrial odor. We are satisfied that the determinative factual conclusions reached by the fact finder rested upon and were fully supported by credible evidence and were not based upon any misuse of the view as evidence.

DRESCO MECHANICAL CONTRACTORS, INC. v. TODD-CEA
531 F.2d 1292 (5th Cir.1976)

After Todd designed and manufactured a boiler and combustion controls for owner/plaintiff, the boiler exploded causing property damage. The owner/plaintiff sued Todd. After the owner prevailed against Todd based on negligence, Todd asserted a claim against Austin. The jury verdict against Austin was set aside by the trial judge and an appeal taken.

Clark, Circuit Judge. Todd's action against Austin rested on a theory of negligent design [by Austin, a consulting engineer retained by the owner who had specified that a dual timer system be used]. To establish a cause of action based on such a theory under Georgia law, Todd was required to establish that Austin's acts or omissions were a breach of duty to observe the standard of care owed to those with whom Austin dealt and that this breach was a proximate cause of the occurrence from which damage was suffered. As an engineering firm, Austin's duty of care was that of a responsible professional person or organization.

[It is] . . . the obligation to exercise a reasonable degree of care, skill, and ability, which generally is taken and considered to be such a degree of care and skill as, under similar conditions and like surrounding circumstances, is ordinarily employed by their respective profession.

[To fulfill its burden to establish the failure to observe this standard] Todd presented the testimony of several witnesses to show that Austin had established the original requirement for a "completely redundant . . . dual flame safeguard system" and had refused to approve a change of a single timer system when Todd's engineers described it to them as "unsafe." [Expert] witness Horton Rucker, an employee of an engineering firm not involved in suit, explained why he believed the explosion to have been caused by the dual timer system. He stated that the dual system presented a "bad situation" because of the possibility that the timers would get out of synchronization and send improper signals to the system. He said he had no personal experience with such a system in his twenty-five years of experience in the field of boiler and burner-control engineering.

[Expert] James Warren of Honeywell, Inc., the manufacturer of the timers, testified that he "[didn't] believe [he] would try using these particular programmers in parallel" because "as long as everything [was] going well they would probably work all right. But if things started going wrong, they would get false signals and get confused."

In short, Todd's proof presents a convincing theory of the *cause* of
the explosion and for the proposition that Austin participated in—or
perhaps even dictated—the decision to install the system whose malfunction lay at the heart of theory. What is missing is any sort of unequivocal statement by any witness that Austin's specification of the
dual timer system or its manner of reviewing Todd's derailed plans
failed to conform to the ordinary accepted practices of the engineering
profession. [The court in critically commenting on counsel's examination of witness Rucker describes it as demonstrating] considerable lack
of communication between counsel and witness, and neither the questions nor the answers establish whether Rucker was describing a "bad
situation" inherent in the design or a "bad situation" that obviously
resulted in this instance. Rucker stated that he was "totally unfamiliar"
with dual timer systems prior to his inspection of the system involved
in the case after the explosion. However, he did not intimate that this
lack of familiarity was based upon the fact that such systems were not
safe to use or not customarily put to use by others. Thus, he could have
no basis for a professional judgment that a dual timer design was inherently unsafe or failed to conform to engineering practice. [The court
noted that witness Warren did not testify] to the professional standard.
It is true that he said he "didn't think" he would use such a system, but
neither his position nor any reasons for it are free from ambiguity.

In marked contrast to these statements is the testimony of
Austin's independent expert witness, Alderman . . . who . . . stated une-
quivocally that the original Austin specifications were "excellent" and
"complied generally and extrapolated on" the generally accepted practice of mechanical engineers in Georgia during the relevant time
period. [Alderman] further stated that the specifications for the dual
flame safeguard system in specific met the accepted standard and that
it was the standard practice for consulting engineers to review builders'
wiring diagrams for general compliance with the specified concept. He
concluded his direct testimony as follows:

Q: Mr. Alderman, will you state whether or not the use of dual timers if
they are properly wired is a practice which can be carried out safely in a
flame safeguard system, if they are properly wired?

A: There's no doubt in my mind that an expert wiring control designer
and system designer could successfully wire dual flame scanners, relays
and timers.

Q: Is the practice of not checking the wiring, details of shop drawings for
workability, is that the general accepted practice among consulting
mechanical engineers in this State and was the practice in 1969, 1970
and 1971?

A: It is the common practice.

In reviewing the correctness of the trial court's decision to take a case from the jury by directed verdict or judgment notwithstanding the verdict, the Court should consider all of the evidence— not just that evidence which supports the non-mover's case—but in light and with all reasonable inference most favorable to the party opposed to the motion.

[I]f there is substantial evidence . . . of such quality and weight that reasonable and fair-minded men in the exercise of impartial judgment might reach different conclusions, the motions should be denied, and the case submitted to the jury . . . . [It] is the function of the jury as the traditional finder of the facts, and not the Court, to weigh conflicting evidence and inferences, and determine the credibility of witnesses. When this test is applied here it requires the conclusion that there was no substantial evidence opposed to the motion on the element of Austin's violation of the standard of care imposed on it. Therefore, the trial court properly refused to allow the jury's verdict to stand.

DISCUSSION QUESTIONS

1. Explain and illustrate the following terms: (1) direct examination, (2) cross examination, (3) hypothetical questions, and (4) refreshing the memory of the witness.

2. Why may an expert witness express an opinion when conclusions and opinions are generally inadmissible in the trial of a case?

3. Discuss the policy questions raised by Federal Rule 740, which now allows an expert witness to testify to the ultimate issue such as in a case of pollution by negligent discharge of waste materials. Is the resolution of difficult technical issues advanced in such cases when expert witnesses for the two parties present conflicting theories on the ultimate issue as to the consequences of a particular chemical pollution in the ground water (e.g., one witness testifies that the substance is the proximate cause of birth defects and the other testifies that there is not causal connection?)

4. Discuss the conditions under which standard text books may be used in examining expert witnesses at a trial.

5. In what kinds of cases may an engineer play an important role as an expert witness?

6. During the discovery period of a lawsuit in which a group of farmers alleged that their well water was being contaminated by chemicals discharged by chemical company X, which was located approximately 5 miles

from the nearest well, Professor Y, who was contacted by the lawyer defending company X, made a study of the facts in expectation of serving as an expert defense witness. However, when Y concluded, based on the strata and ground water conditions, that X was a contributor to the contamination of the wells, the attorney paid Y's agreed compensation and advised Y that he might be asked to conduct some additional studies but for now no more of his services would be needed. When the case goes to trial, Y learns that Professor Z has testified for company X that there is no connection between the pollution in the wells and the company. Does X have any ethical or legal obligations in this case and, if so, what?

7. Discuss the issues raised by an agreement to compensate an expert witness on a contingent fee basis.

8. Should the courts and administrative agencies promulgate written instructions, admonitions and warnings for the expert witness who is to testify in a proceeding before a court or administrative agency and what sanctions should be imposed?

INDEX